D.E. Salmon

Special Report on the Cause and Prevention of Swine Plague

D.E. Salmon

Special Report on the Cause and Prevention of Swine Plague

ISBN/EAN: 9783337329143

Printed in Europe, USA, Canada, Australia, Japan

Cover: Foto ©berggeist007 / pixelio.de

More available books at **www.hansebooks.com**

U. S. DEPARTMENT OF AGRICULTURE.

BUREAU OF ANIMAL INDUSTRY.

SPECIAL REPORT

ON THE

CAUSE AND PREVENTION OF SWINE PLAGUE.

RESULTS OF EXPERIMENTS

CONDUCTED UNDER THE DIRECTION OF DR. D. E. SALMON,
CHIEF OF THE BUREAU OF ANIMAL INDUSTRY.

BY

THEOBALD SMITH, Ph. B., M. D.

PUBLISHED BY AUTHORITY OF THE SECRETARY OF AGRICULTURE.

WASHINGTON:
GOVERNMENT PRINTING OFFICE.
1891.

TABLE OF CONTENTS.

LIST OF ILLUSTRATIONS.

5

WASHINGTON, D. C., *July* 20, 1891.

SIR: I have the honor to submit herewith the second special report upon the investigations of infectious swine diseases. The first report dealt mainly with hog cholera, while the accompanying report is devoted to swine plague.

In this volume the investigations are reported in detail which have led up to a separation of swine plague as an independent disease from hog cholera. The difficulties surrounding investigations of infectious swine diseases have been very much increased by the frequent association of both hog cholera and swine plague in the same herd. This has necessitated frequent repetitions of investigations before positive results could be obtained.

The importance of thoroughly scientific investigations of infectious animal diseases is becoming more and more apparent, and, owing to the commercial interests at stake, the subject is now assuming an international character. Even though investigations of this nature do not at once suggest the means of dealing satisfactorily with such diseases, it is, nevertheless, essential that we should know as much as possible of the nature of every one of them within our boundaries if we expect to bring to bear upon them all the knowledge which is accumulating so rapidly in various parts of the civilized world. The investigations in this report prove the existence of a virulent swine disease due to specific bacteria and its identity with the disease of *Schweineseuche*, prevalent in Germany. It also demonstrates the wide distribution of this class of bacteria, chiefly as varieties of feeble disease-producing power in the upper air passages of various domesticated animals. It is not improbable that this group of bacteria may, when exceptionally virulent, attack other domesticated animals, and indeed we know that the bacteria of at least one form of fowl cholera can not be distinguished from those of swine plague by the bacteriological methods now in use. In short, this group of bacteria which appears to have a world-wide distribution, may be considered as very dangerous to our domesticated animals, and, consequently, the care with which they have been studied, more particularly with reference to their disease-producing properties, is fully justified.

With the publication of this and the preceding volume the information is given by which a diagnosis can be made of the two principal

7

infectious diseases to which our swine are subject. The nature and cause of these diseases are made clear, and knowing these, we can apply the curative and preventive measures which the progress of medical science has made possible, or which future researches may show to be available.

D. E. SALMON,
Chief of the Bureau of Animal Industry.

Hon. J. M. RUSK,
Secretary of Agriculture.

INVESTIGATIONS CONCERNING THE NATURE AND ETIOLOGY OF SWINE PLAGUE.

INTRODUCTORY.

Since 1886 the bacteriological investigations carried on in the laboratory of the Bureau of Animal Industry on infectious swine diseases have revealed the existence of pathogenic bacteria differing markedly from those causing hog cholera. They preferably attack the chest organs of swine, the lungs, and serous membranes covering these vital organs. The difficulty encountered from the beginning in obtaining a clear understanding of the action of these disease germs is the frequent mingling of hog cholera with this disease, which has been called swine plague. In some outbreaks only hog-cholera bacilli are detected, in others only swine-plague bacteria, while in the majority of outbreaks thus far studied both bacteria are associated together. This condition is very likely due to their wide distribution over the country.

The importance of swine-plague bacteria is not limited to swine diseases alone, for they belong to a group of bacteria which in various parts of the world have attacked other species, such as cattle, horses, game, and fowls, and the question whether there may not be a transmission of such diseases from one species of farm animals to another must not be lost sight of.

The swine-plague bacteria have been very carefully studied in view of their importance, and attention has been given more directly to their pathogenic effect on the smaller experimental animals, such as mice, rabbits, guinea-pigs, pigeons, and fowls, in order that a proper understanding of their relative virulence could be obtained. Investigations in various diseases have brought to light bacteria which, although identical so far as we are able to determine, vary greatly in their virulence, and it seems now that the relative virulence or disease-producing power must be looked upon as the chief criterion of danger. Feeble virulence signifies a feebly infectious localized disease. A high degree of virulence signifies the power to spread and perhaps to attack other species. This important factor of virulence can only be measured by the inoculation of animals of various degrees of susceptibility. The great variation in the virulence of both swine-plague and hog-cholera bacilli has been demonstrated in this way, and when we come down to the lowest grades the problem arises, can such varieties produce disease at all, and are they not merely accidentally present? An answer to this question can not be given categorically at present, and its solution will demand continued, undivided attention to this subject of disease germs for some time to come.

Another fact, the significance of which has been discussed in the text, is the existence of bacteria in the air passages of various domesticated animals during health, which are not distinguishable from swine-plague bacteria excepting by feebler pathogenic effect.

The question whether swine plague is an infectious disease by itself, demanding the attention of the veterinarian, the agriculturist, and the Government, was practically solved, though not absolutely demonstrated, by the investigations of 1887. The repeated attacks upon this work made it desirable to spend much more time than was actually necessary in the investigation of outbreaks as they came within reach of facilities from time to time, and now a large amount of material has been collected which fully confirms the position taken in 1887, that there are at least two swine diseases of an infectious character. These have very likely existed together for a long time, but a differentiation could only be effected by the advanced position of bacteriology and the ample means provided by the Government.

The frequent association of hog-cholera bacilli with swine-plague bacteria has made it necessary to pay some attention to hog cholera as well. These investigations are based on the facts already published on hog cholera and the bacillus of that disease in the special report on hog cholera (1889). This report is therefore necessary to an understanding of the contents of the present volume so far as the strictly experimental portion is concerned.

It is to be hoped that these investigations will prove not only of advantage in the protection of swine from disease, but will be a basis for the investigation of diseases of other domesticated animals due to this group of bacteria, which may appear at any time owing to the growing complexity of intercourse between different sections of the country, modifications in the methods of stock-raising, etc.

It must not be assumed that our knowledge of the etiology of swine diseases is complete even with the advances made thus far in swine plague and hog cholera. There are important questions with reference to both diseases which demand elucidation. Especially is it desirable to investigate those outbreaks more carefully in which we find the appearances of hog cholera without the bacilli in the internal organs. The complete mastery over a disease is only to be obtained by complete insight into its causes. If all the suggestions derived from scientific investigations, conscientiously applied, fail to prevent the appearance of a given disease it signifies that our information is as yet incomplete, and that there are some still unknown channels through which the bacteria of such disease are distributed.

To devise a proper plan for an intelligible presentation of the investigations and results which will be satisfactory to those who are not immediately interested in the investigations, and simply desire to know the deductions so as to be guided by them, has been a difficult one, and has demanded the occasional repetition of statements. In general all those facts not necessary to an understanding of the text have been put into smaller type. All the investigations proper, including a detailed description of the various outbreaks, autopsy and bacteriological notes, test inoculations of swine and smaller animals, have been kept together under the respective outbreaks, while all essential information derived from these investigations and experiments is to be found in the subsequent sections of the report.

The contents, therefore, have been arranged in the following order:

The methods of work are first described, followed by a brief review of earlier and a more detailed account of later investigations. Then come a general description of the swine-plague bacteria, of their disease-producing power, and some facts showing the wide distribution of this group of bacteria among domesticated animals in a healthy condition. A chapter is also added, giving a brief account of the investigation of others in this field. Lastly, all the results and practical observations which may be of value to the farmer are brought together.

The writer acknowledges the continued assistance since 1886 of Dr. F. L. Kilborne, veterinarian of the Experiment Station, who has had charge of the experimental animals, has assisted in many of the post mortem examinations, and has performed the inoculations upon swine. In the laboratory, Dr. V. A. Moore has assisted, since 1888, in general pathological and bacteriological work. The special services of these gentlemen are also referred to in the text.

BRIEF DESCRIPTION OF THE METHODS EMPLOYED IN THE INVESTIGATIONS.

The difficulties that are met with in the investigation of infectious swine diseases are not those usually encountered in bacteriological investigations. In the first place outbreaks occur as a rule some distance from a laboratory. Yet the amount of bacteriological work demanded by each case, in order that any conclusion may be reached, can not be done excepting in a well equipped laboratory. Hence it has been our effort to transfer, if possible, cases of the disease to the Experiment Station, and also to keep up the disease there by exposing fresh cases. In fact, only those investigations carried on in this way can be regarded as complete, while cultivations made in the field are at best restricted to a few or only one organ, and plate cultures are out of the question.

Another difficulty is due to the belated information which we obtain of any given outbreak. Usually from one-third to two-thirds of the herd have perished before the investigations are begun. This is unfortunate in swine plague, because the earliest cases are the most satisfactory from a bacteriological standpoint, since the swine-plague bacteria are most easily demonstrated in such cases. It seems that the infection passes through a herd quite rapidly, and those animals which live longest have reached a stage of the disease in which the swine-plague bacteria may have entirely disappeared.

A third difficulty to be contended with is the rapid death of the animals, which makes a thorough examination of each animal impossible. The alternative before us is either to limit our examination to a certain number of organs in every animal, or else to examine a few very thoroughly. While we have endeavored to meet both demands, of examining every animal, and also doing it as thoroughly as possible, we have but partially succeeded in this, owing to lack of facilities and assistance. The pathological changes differ so much from animal to animal, the extent of the organs involved varies to such a degree, that there is a great temptation to examine every animal in the hope that most information can be obtained in this way.

The investigations consisted in watching the course of the disease in the infected animals, in post-mortem examinations of those that died, in microscopic examination of the diseased tissues, fresh and hardened, and in bacteriological examination of a certain number of organs. The post-mortem examination included the various organs and tissues, with the exception of the brain and spinal cord, which were exposed in a very few cases only. The skin was first reflected from the thorax and abdomen, the abdomen carefully opened, and the spleen or a portion of it first removed with sterile instruments and reserved for further examination. From any fluid or exudate present a minute quantity was transferred to culture media by means of a platinum wire bent into a loop about three to four millimetres in diameter and soldered into a glass rod. The thorax was next opened by removing the sternum, and from any effusion or exudate cultures were made at once. The various organs destined to be used for bacteriological examination were removed in whole or in part before being soiled.

Portions of organs designed for microscopical examination were in part examined fresh by making sections with the razor, teasing, etc., in part hardened in 95 per cent. alcohol and infiltrated with paraffin before the sections were made. These were stained in different ways, chiefly in alum carmine and in alkaline methylene blue (Löffler), which was found most satisfactory in bringing out the swine-plague bacteria.

The various cultures were made in the following manner: With a platinum spatula a small area of the surface of an organ was thoroughly scorched, and from the scorched area minute particles of tissue were cut out with flamed scissors and forceps, the scorched layer being rejected. These particles were then transferred to various media for tube and plate cultures and also used for the inoculation of animals. Another mode of inoculating cultures used was to pierce the scorched area with a straight stiff platinum wire, which was then stirred about in liquefied media or rubbed over the surface of agar. In this way only a minute trace of the parenchyma could be transferred, and it was only used when bacteria were numerous enough to be detected under the microscope. In the notes given with the various outbreaks the platinum loop and wire mean respectively the looped wire and the straight wire. Note is merely made of this to indicate the difference in the quantity of material used to inoculate the culture media.

The more usual method in vogue of disinfecting the external surface of organs by placing them in a 5 per cent. solution of carbolic acid or a one-tenth per cent. solution of mercuric chloride for a short time, then making several incisions in planes at right angles to one another into the depth of the organ, was not employed, because the method already mentioned of scorching a small portion of the surface seemed safer and more satisfactory. When cultures were inoculated with blood this was done from the right ventricle while the heart was still *in situ.* The thin ventricular wall was scorched over a small area, then the edge of the spatula was used to burn through. The platinum wire or loop, or the glass pipette, if blood was to be collected, was inserted through the hole thus made. The same methods were employed in the examination of the small experimental animals.

For the cultivation of swine-plague bacteria agar-agar as usually prepared with bouillon, peptone, and common salt, and the ordinary peptone bouillon, are the best media. In the earlier investigations nutrient gelatine was chiefly employed, but it was found that some varieties of swine-plague bacteria may refuse to multiply in it, hence it was discarded. In addition to plate cultures for the isolation of these bacteria the inclined surface of agar in tubes is of great service. In many cultures the colonies appeared completely isolated on the agar surface, thus enabling us to start from single colonies when plate cultures had not been prepared at the same time.

For the inoculation of smaller animals with bacteria from cultures two methods were employed: (1) A small pocket was made in the sub-cutis by an incision through the skin with sterilized scissors and a loop of the growth rubbed into this pocket; (2) liquid cultures were prepared by inoculating peptone bouillon and allowing the liquid to become clouded in the thermostat or by transferring the growth from agar tubes into sterile bouillon or water and making a suspension. The liquid was then injected with a hypodermic syringe into the sub-cutaneous tissue or into a vein. The hypodermic syringes used were of the ordinary pattern, since the various others devised, such as the Koch syringe and the asbestos-packed syringe, were found leaky and ineffi-cient and did not enable us to inject a definite quantity with accuracy. In this investigation these syringes could the more easily be dispensed with, since both swine-plague and hog-cholera bacteria are sporeless and killed by disinfectants in very dilute solutions. The syringes and needles were disinfected both before and after use by filling them with

5 per cent. carbolic acid and allowing this to remain for 15 or 20 minutes. The acid was then removed with boiling water. Great care was taken that no fluid passed beyond the piston. If it did the syringe was taken apart and the parts immersed in 5 per cent. carbolic acid. While this method of making injections may seem theoretically objectionable, it has nevertheless at no time proved the source of any accidental infection. This method is not applicable, however, to bacteria which produce spores, or which prove very resistant to disinfectants, such as anthrax and tuberculosis. In general syringes were dispensed with and the first method employed when the accurate measurement of the quantity of virus to be injected was not essential.

The technical difficulties surrounding the problem of the causation of swine diseases are mainly due to the intermingling of two diseases, hog cholera and swine plague. As this problem will be discussed further on, only those points need be considered here which involve the methods employed. The important question to be decided in every outbreak is whether one or both kinds of bacteria are present. This can only be determined by bacteriological investigation. It is evident that to examine every organ of every pig bacteriologically is a task of enormous dimensions, hence the simplest reliable method must be employed. At first thought it would seem sufficient to make plate cultures from the various organs and examine the colonies as they develop. There are, however, objections to this method taken alone. Certain varieties of swine plague bacteria, especially the most virulent, do not as a rule grow in gelatine, while on agar the colonies of hog cholera can not be distinguished positively from those of swine plague, unless the bacteria composing these colonies are examined. For a rapid and reliable determination of the presence or absence of these two kinds of bacteria. I have found in addition agar and bouillon tube cultures necessary. The motile hog cholera bacilli are detected at once in the bouillon and in the condensation water of the inclined agar cultures. Such cultures may then be plated, and fresh cultures of the hog-cholera bacilli and the swine-plague bacteria obtained by inoculating from isolated colonies. These cultures are then in condition to be tested on smaller animals to make the diagnosis complete. While it is desirable that when both kinds of bacteria are present they should be isolated from as many animals as possible, it is obvious that the presence of virulent hog-cholera bacilli in a single animal of a herd leads to the inference that they are most likely present in all the remaining diseased animals.

The detection of swine-plague bacteria in liquid cultures containing hog-cholera bacilli is, as a rule, not difficult. The best method consists in the examination of the liquid unstained in a drop suspended from the under surface of a cover-glass resting on the margins of a cell on or in the slide. The most convenient cell I have found to be a thin glass ring fastened to the slide with Canada balsam. A little glycerin is placed on the ring so that when the cover glass with the drop on it is inverted and laid on the ring the immersion lens will not draw it up.

The hog-cholera bacilli are readily detected in this drop by their motion.* If the border of the drop be examined, the two kinds of bacteria may then be distinguished by a slight difference in size in favor of the hog-cholera bacilli. When the growth from the condensation water of agar cultures is to be examined, it must be diluted with bouillon or sterile water kept preferably in tubes, just as the bouillon is, in small quantities. The examination of cultures in this manner can not be dispensed with in the study of these diseases. When the presence of one or both kinds of bacteria has been positively determined, I find staining of very little value in subsequent work and simply examine fluids as described. When colonies are to be examined a drop of bouillon or water in which a trace of growth is stirred up is, of course, necessary.

Another advantage of the bouillon and the condensation water of agar tubes rests upon the fact that motile bacteria do not always become active at once when transferred from agar or gelatine into liquids for microscopical examination. Hence their motility may be entirely overlooked unless they are examined after having multiplied in liquids for some hours at least.

It should likewise be borne in mind that the colonies of a few bacteria in a bit of tissue or drop of effusion from the diseased body may escape attention on a plate when many other colonies are present, and that they may be entirely lost on the second or third plate. In bouillon, however, both kinds of bacteria have for a time the same opportunities for multiplication, and both may be detected on the following day, though there may have been originally a great difference in the numbers introduced. While agar plates have occasionally failed to demonstrate the presence of certain bacteria, owing either to a failure to multiply or to the rapid growth of other bacteria, or to rapid drying out of the agar layer, bouillon and agar tube cultures made at the same time have furnished the desired information.

In order to detect both kinds of bacteria it is therefore essential that bouillon and inclined agar in tubes be used with agar plates. The method pursued in the investigations detailed in full in this report was always to prepare plates from lungs. When there was no time for the preparation of plates from other organs, such as spleen, liver, and kidneys, tubes of bouillon and agar were used in place of them, and, if necessary, plates were prepared from these.

A minor difficulty, but one which may prove of more serious consequence to beginners, is the frequent encountering of bacteria other than those producing the disease in the organs of diseased swine. A perusal of the bacteriological observations in this report will show how much time has to be spent in isolating these bacteria and determin-

---

* See Special Report on Hog Cholera, 1889, for the characters of these bacilli.

ing what relation they bear to the disease. Many of them can eventually be traced to the intestines where they commonly vegetate. Their presence in the internal organs may be accounted for by the extensive lesions in lungs and intestines which serve as entrances into the blood. This presence of strange bacteria has also been observed in other infectious diseases by other observers, and attention has been called to it in connection with hog cholera in the special report on that malady, p. 58.

The detection of these foreign bacteria has been in large part due to the use of bouillon. As they are usually present in very small numbers their colonies on plates might easily be overlooked or attributed to accidental infection. They would appear only on the first plate among large numbers of other colonies and most likely be overlooked, since the development of colonies on crowded plates is limited, and they frequently fail to show any special features unless they have time and opportunity to expand. This is true of gelatine and particularly so of agar, on which colonies show, at best, few differences. When bouillon is inoculated all bacteria have for a time equal opportunities to develop, as stated above, and they at once thrust themselves upon our attention. Such mixed cultures have been and are still styled contaminated by those who fail to see their real significance. Such criticisms may, however, be safely left to take care of themselves at the present time.

As an illustration of the foregoing, I have frequently found in the bottom of bouillon tubes into which a bit of tissue had been introduced large spore-bearing bacilli, which have been referred to in the various reports as "anaërobic," "post-mortem" bacilli. These bacilli do not develop in fresh cultures, nor would they appear on plates. The bottom of the culture liquid and the bit of tissue furnish for the time a suitable soil.*

The use of animals in the isolation of bacteria is of great value in swine diseases. Rabbits are peculiarly susceptible to swine-plague bacteria. Inoculated with the more virulent varieties they die within 20 hours, and the inoculated bacteria can be obtained from the various organs. Frequently when cultures failed to determine the presence of these bacteria in tissues rabbit inoculation was still successful. When a mixed outbreak is under examination it is evident that since rabbits do not succumb to minute doses of hog-cholera bacilli in less than 7 days, the rabbits will die of swine plague first and the hog-cholera bacilli will not be obtained from their organs. There are, however, attenuated varieties of swine-plague bacteria frequently encountered in mixed outbreaks which prove fatal to rabbits in from 3 to 10 days. In these cases when pure cultures are inoculated there is more or less suppurative destruction of the subcutis starting from the point of inoculation, while the internal organs are quite free from changes and bacteria also. In such protracted cases, provided both kinds of bacte-

---

* For a simple method of cultivating such bacteria in addition to those now in use for anaërobes, see foot-note, p. 21.

ria were originally present in the tissue inoculated, both kinds may appear in the rabbit cultures, or only the hog-cholera bacilli may have become disseminated through the body, while the swine-plague bacteria may be limited to the inoculated locality. Since hog-cholera bacilli do not produce much local inflammation, whenever this is extensive in rabbits inoculated directly from the pig, it is pretty certain that swine-plague bacteria and perhaps other still unknown bacteria may have caused it and should be looked for.

Rabbit inoculation may thus prove very serviceable, but the post-mortem appearances must be carefully interpreted in connection with the bacteriological observations. In any case they rarely bring to light hog-cholera bacilli when the inoculated animal succumbs to swine plague before the fourth or fifth day.

A few words are necessary to define some of the anatomical terms used in the autopsy notes, inasmuch as a thorough description of the visceral anatomy of the domesticated animals has not yet been made. Since the lesions are confined chiefly to the lungs and intestines our remarks will be confined to these organs.

When inflated through the trachea after the sternum is removed, and while the lungs are still in their natural position in the thoracic cavity, it will be observed that the surface resting against the ribs laterally is the most extensive. That surface resting upon the diaphragm comes next, while the ventral aspect is the smallest. (See Plates I, II.) The right lung is made up of four lobes, the left has only three. (In text-books on anatomy the left lung is considered as being made up of only two.) In both there is a large principal lobe resting upon the diaphragm and against the adjacent thoracic wall. This lobe forms the major part of each lung. The remainder, occupying the anterior (or cephalic) portion of the cavity, is made up of two small lobes, one extending ventrally (or downward in the standing position of the animal) and in the expanded state covering the heart laterally, the other extending towards the head and overlapping the base of the heart. These small lobes may be denominated the ventral and cephalic lobes, respectively. The right cephalic lobe is longer and more distinct from the ventral lobe than the corresponding left cephalic. Wedged in between the two principal lobes and resting on the diaphragm is a small lobe, pyramidal, belonging to the right lung (azygos or median lobe). This lobe rests on the left, against the mediastinal membrane, and on the right it is separated from the right principal lobe by a fold of the pleura passing from the ventral thoracic wall to inclose the inferior vena cava. This small lobe is almost completely shut off, therefore, from the other lobes by folds of the pleura.

When the trachea and its branches have been examined it is easier to understand this division into lobes. The trachea divides in the thorax into two principal branches or bronchi. These bronchi pass into the principal lobes, straight to the caudal border, giving off a number of small branches along their course. Very near the place of bifurcation the left bronchus gives off a large branch, which ramifies in the substance of the left ventral lobe. From this branch another goes to the cephalic lobe. In some lungs the branches of these two lobes arise together from a very short, scarcely perceptible trunk, and are of nearly equal size. The bronchial supply of the right lung differs materially from that of the left. About 2 centimetres from the bifurcation the trachea gives off a small bronchus, which supplies the right cephalic lobe exclusively. At the bifurcation the right bronchus sends a short branch to the small median or azygos lobe and one to the ventral lobe.

The major portion of the large intestine of the pig consists of a duplicature which is coiled upon itself somewhat like a watch spring. The cæcum is bound down to the dorsal wall of the abdominal cavity on the right side. Similarly the rectum and a small portion of adjacent colon is fastened down by peritoneal folds. If, starting from cæcum and rectum as fixed points, the entire large intestine be folded on itself at the middle point of its length and the whole coiled up, beginning at this point as the free central end, we have imitated the general arrangement of the large intestine in the abdomen. The coiled part is loosely bound together and readily movable in the abdomen. As regards the dimensions the following measurements are taken from a male pig about 7 months old, and 39 inches long from the tip of snout to root of tail. The intestine was distented with normal feces : 1, cæcum, extending from closed end to ileo-cæcal valve, 7½ inches long; 2, from valve to duplicature or bend in the center of the coil, 5 feet ; 3, from the latter point to anus, 6 feet. It will be noticed that the folding takes place in the middle of the entire length of the large intestine.

In the description of the regions the duplicature is a convenient point at which to distinguish the upper from the lower colon. Where the rectum begins it would be difficult to determine without careful anatomical observations. Provisionally, its limit if determined by its old original signification may be placed at eight to twelve inches from the external opening or anus.

In examining the large intestine it is best to begin the separation at the fold or bend in the center of the coil and continue until the cæcum and lower colon are reached. The cæcum is easily removed with the small intestine attached. The lower colon is associated with the duodenum in the same mesentery and is crossed by it. Care must be taken not to injure either tube at this point in attempting to separate them. In beginning it is best to tie a string around the tube at the flexure as a landmark for future reference.

The writer has used, apparently without discrimination, the English and the metric system of weights and measures. It will be noticed, however, that in the more strictly scientific part of the work only the metric system is used, while in those portions which may be interesting to a wider circle the English system is substituted.

I.

During the summer of 1886 the writer made a number of post-mortem examinations of diseased pigs in Illinois,† for the purpose of identifying, if possible, the swine diseases existing in the Western States with the disease of hog cholera (or swine plague as it was then denominated), under observation since November, 1885, on the Experiment Station near Washington. It will be remembered that the hog-cholera bacillus was discovered in November of 1885, and carefully studied as to its morphological and pathogenic characters.

The post-mortem examinations were made in the field, the spleen removed, placed in a sterilized bottle, and the cultures made either immediately or several hours later, indoors. During the first visit to Illinois in July, 1886, pigs were examined in Marion, Champaign, and Henry Counties. In some cases the lungs were diseased. In most cases the intestines were more or less ecchymosed, while ulcers or indurations of the mucous membrane were rare. What was most surprising, however, was the absence of any bacteria in the spleen of seven out of eight animals examined. In the eighth case there was considerable pneumonia associated with pigmentation of the mucosa in the large intestines. The culture from the spleen contained two species of bacteria, the *bacillus coli communis*, a common inhabitant of the digestive tract of the pig and other domesticated animals, and a small, non-motile bacterium which possessed pathogenic properties as the inoculations made at that time distinctly demonstrated. These bacteria were isolated by inoculating a rabbit subcutaneously with a bouillon culture containing both kinds of bacteria. The rabbit died in 7 days with extensive inflammation of the subcutaneous tissue starting from the point of inoculation. Cultures from the spleen, liver, and blood contained only one form, the oval non-motile bacterium. A second rabbit, which received one-sixth cubic centimetre of a pure bouillon culture of these bacteria under the skin, died in 3 days with beginning peritonitis. Two mice, inoculated in the same manner, died respectively 1 and 2 days after inoculation.

---

*Already published in detail in various reports of the Department of Agriculture.
†See Report of the Bureau of Animal Industry for 1886, p. 76.

Much time was spent in studying the motile bacillus, owing to its superficial resemblance to hog-cholera bacilli. Any one familiar with intestinal bacteria will, I think, at once concede, after reading the description on page 78 of the report referred to, that this was the *bacillus coli*. The peculiar expanding and rapid growth on gelatine, the coagulation of milk, the offensive, putrescent odor of the cultures, are properties which do not belong to hog-cholera bacilli. Moreover there was no manifestation of any pathogenic power. The small non-motile bacteria were identical with those named throughout this report as swine-plague bacteria.

II.

Several months later the writer, having found the information obtained on this journey quite different from what was expected, undertook another journey in the same State.* As usual the various reports of swine disease found in the papers vanished before the writer's approach, and after much search a herd was found near Sodorus, Champaign County, in which a number of animals had already perished and some were very sick. Two of the latter were killed. In No. 1 there were extensive pneumonia and some large ulcers in the large intestine. In No. 2 there were the same pneumonia, a very large spleen, and one ulcer on the valve in the large intestine. Various cultures were made at the time and blood was collected in pipettes which were sealed in the flame. The result of the bacteriological examination was briefly as follows:

From No. 1: The spleen contained only hog-cholera bacteria. The blood (first pipette) hog-cholera bacteria and streptococci. The blood (second pipette) contained only swine-plague bacteria.

From No. 2: The spleen, blood, and pleural cavity each contained swine-plague bacteria.

Sections of the lung tissue of No. 2, hardened in alcohol, showed large numbers of swine-plague bacteria in the alveoli.

With swine-plague bacteria from the pleural tube cultures of pig No. 2 a considerable number of inoculations were made upon smaller animals to determine their pathogenic power as compared with the swine-plague bacteria already found. The inoculations were made by injecting definite quantities of bouillon cultures which had been inoculated from single colonies on gelatine plates.

One mouse, one-eighth cubic centimetre subcutaneously, dies in 45 hours.
One mouse, three-sixteenths cubic centimetre subcutaneously, dies in 24 hours.
One mouse, one-twelfth cubic centimetre subcutaneously, dies in 2 days.
One mouse, one-twelfth cubic centimetre subcutaneously, dies in 6 days.

In these animals there were no marked lesions. In some the bacteria injected were present in large numbers in the various organs; in others they were very scarce.

---

*L. c., p. 79.

One rabbit, one-eighth cubic centimetre subcutaneously, dies in 4 days.
One rabbit, one-fourth cubic centimetre subcutaneously, dies in 5 days.
One rabbit, one-eighth cubic centimetre subcutaneously, dies in 3 days.

In the rabbits at the place of inoculation the subcutis was much thickened by purulent infiltration, and there was more or less hemorrhagic and exudative peritonitis. In the exudate the injected bacteria were very abundant.

One pigeon, one-half cubic centimetre subcutaneously, dies in 2 days.
One pigeon, three-fourths cubic centimetre subcutaneously, dies in 4 days.

At the place of the inoculation the muscular tissue more or less necrosed. Bacteria not detected in the internal organs.

One guinea-pig, one-fourths cubic centimetre subcutaneously, dies in 6 days with slight local changes and exudative peritonitis and pleuritis.
One fowl, one-half cubic centimetre subcutaneously, dies in 5 days.
One fowl, 1 cubic centimetre subcutaneously, recovered.
In both there was extensive necrosis of the pectoral muscle at the place of inoculation.

The following animals were inoculated by rubbing into the subcutaneous tissue through an incision a loop dipped into a gelatine culture:

One guinea-pig (abdomen) dies in 5 days; subcutaneous lesion very extensive.
One guinea-pig (abdomen) dies in 8 days; subcutaneous lesion very extensive.
One mouse (root of tail) dies in 3 days.
One mouse (root of tail) dies in 4 days.
One rabbit (ear) dies in 9 days; extensive subcutaneous lesions.
Two pigeons (breast) remain well.
One fowl (breast) remains well.

From cultures of the swine-plague bacteria derived from the blood of pig No. 1 two rabbits were inoculated to see if the bacteria from both cases were identical.

One rabbit received one twenty-fourth cubic centimetre bouillon culture; died in 3 days.
One rabbit received a loop of gelatine culture; died in 10 days.
In both there was much local inflammation in the subcutis and peritonitis.

With cultures of the swine-plague bacteria from these two sources a number of pigs were inoculated as indicated in the table:

Pigs.	Date of in- oculation.	Quantity injected.	Source of culture.	Remarks.
		c. c.		
No. 287	Sept. 11, 1886	*2½	Geneseo, Ill............	Dies in 18 days.
No. 289	...do	*2½	....do	Dies in 10 days; generalized jaundice.
No. 363	Oct. 30, 1886	*5	....do	Dies in 58 days.
No. 367	....do	1½	....do	No effect
No. 330	Oct. 4, 1886	*5	Sodorus, Ill	Dies in 9 days; jaundice.
No. 331	....do	*3	....do	Dies in 35 days.
No. 364	Oct. 30, 1886	*5	....do	Dies in 8 days; jaundice.
No. 366	....do	1½	....do	No effect.
No. 374	Nov. 18, 1886	*5	....do	Dies in 11 days; jaundice.
No. 375	... do	*5	....do	Dies in 7 days; jaundice.

* Subcutaneous.   † Into right lung.

It will be seen that of these ten pigs eight were inoculated subcutaneously with doses ranging from $2\frac{1}{2}$ to 5 cubic centimetres of a bouillon culture. Two died from 1 to 2 months after inoculation, the remaining six in from 7 to 18 days thereafter with a peculiar disease of the liver and generalized jaundice.

Two inoculated into the lung tissue through the right chest wall remained unaffected.

The liver in these cases was greatly enlarged and so firm that when removed from the body there was no change of form. There was no obstruction to the flow of bile in the bile ducts. The disease of the liver tissue itself was shown in sections of cases 289 and 375 to be due to inflammatory foci within the lobules. In some of these foci the parenchyma cells were still visible, but very feebly stained ; the trabecular arrangement was destroyed; the nuclei of the cells very much shriveled or absent. In other foci the parenchyma was replaced by numerous round cells. From these observations it would seem that the disease consisted of necrosis of a mass of hepatic cells followed by round cell infiltration. These foci varied in size from one-eighth to one-half the area of the lobule. Almost every lobule was affected, either several small areas or one large area of disease being present in each, situated usually near the periphery. In No. 375 there was also very extensive cellular infiltration beneath the capsule.*

That the disease was induced by the inoculation can not be very well denied in view of the facts in the case. The animals used were from four different lots purchased from four different owners.† No other animal in these lots died with these lesions of the liver. The inoculations must, therefore, be considered as the direct cause of the fatal result.

These experiments are in so far remarkable as future subcutaneous inoculations with swine-plague bacteria from other sources, with one exception, produced no effect. I have not observed this disease since the time these experiments were made. The only explanation which can be made is that these varieties of swine-plague bacteria had a slightly different pathogenic power which manifested itself in the manner described.

* Much ill-considered criticism has been leveled at these experiments and results by F. S. Billings, and they are, therefore, presented again simply to show the unfounded character of these criticisms. While the term cirrhosis used in the earlier reports may not have exactly expressed the diseased condition of the liver, it should also be remembered that the field of comparative pathology is not sufficiently developed to aid us in choosing terms or in finding certain diseases already described and named.

† Nos. 287-294, bought June 1, 1886, when 8 weeks old, from Mr. M. ; Nos. 329-335, bought Aug. 27, 1886, when 8 weeks old. from Mr. B. ; Nos. 363-367, bought Oct. 18, 1886, when 8 weeks old, from Mr. J. ; Nos. 368-382, bought Oct. 18, 1886, when 8 weeks old, from Mr. J. F.

III.

Toward the close of 1886* Dr. Paaren sent to the Bureau portions of lungs from Cerro Gordo County, Iowa. These portions indicated consolidation of the lung tissue with necrotic foci and exudative pleuritis. By inoculation into mice and rabbits swine-plague bacteria were obtained, which, reinoculated in pure culture, produced death within 24 hours. The bacteria, very abundant in the internal organs, showed distinctly the polar stain.† They did not differ in any respect from the bacteria found in Illinois, excepting in their greater virulence on the smaller experimental animals. Two pigs inoculated subcutaneously with 5 cubic centimetres each did not manifest any signs of disease.

IV.

In February of 1887 an outbreak of swine disease came under observation which showed conclusively the transmissibility of the lung disease and the bacteria which are the cause.

The disease was associated in every case with pneumonia and pleuritis of a more or less severe character. In the course of the disease caseation of the involved lung tissue was frequently observed. Intestinal lesions of a kind hitherto observed but once before‡ were present in the earlier cases. These investigations have been criticised by F. S. Billings, if his statements deserve the name of criticism, because, in addition to the swine-plague bacteria found in almost every case, hog-cholera bacilli were detected in the later cases. It is difficult to discover from his statements which of the two bacteria he objects to. His own report leaves us wholly in the dark which kind he himself has found in Nebraska, owing to the ambiguous manner in which the bacteria found by him have been described. We may therefore pass over these criticisms and briefly summarize the investigations§ which have since been confirmed by much more extended ones.

On January 31 a small number of pigs were purchased from a farm adjoining the Experiment Station for the purpose of testing the effect of different cathartics on the healthy pig. The animals had been purchased from this farm because no disease had existed there for several years. These pigs (Nos. 402–406, inclusive) were put in a pen by themselves. On the following day one (No. 406) was found dead with extensive ulceration of the large intestine, but no lung disease. This occurrence of course spoiled the entire experiment as planned. The disease was regarded as hog cholera by me, although I did not examine the

<hr/>

* L. c., p. 92.
† For the meaning of this expression, see p. 85.
‡ Report of the Bureau for 1886, p. 66.
§ For a full account of the autopsy notes and bacteriological examination, see report of the Bureau for 1887, p. 86.

animal or make cultures, being engaged in other work at the time. Another fact which led to the supposition that this disease was different from the disease which appeared later in the other animals of this lot was told by the owner. Nos. 405 and 406 were the only pigs which had not been confined in pens, and which, therefore, may have been exposed to infection on the farm which did not reach the penned animals.

At the time No. 406 died (February 1), the other animals appeared well. Some of these were distributed into small pens by themselves and fed different doses of cathartics. Meanwhile no other animal of this lot died until 15 days later, when 403 died. The following table gives information concerning the rest:

Pigs.	Date of death.	Remarks.
No. 406	Feb. 1....	Intestinal ulcers.
No. 403	Feb. 16...	Ventral lobes of lungs diseased ; croupous inflammation of large intestine.
No. 405	Feb. 18...	Extensive pneumonia and pleuritis ; croupous inflammation of large intestine.
No. 402	Feb. 19 ..	Slight pneumonia ; croupous inflammation of large intestine.
No. 404	....	Not affected.

As regards the bacteriological examination the following may be said:

No. 406. None made because disease supposed to be hog cholera.

No. 403. None made for same reason.

No. 405. Swine-plague bacteria found in lungs; other organs not examined.

No. 402. Bouillon cultures made from pleural shreds, spleen, liver, and blood. Gelatine cultures made from blood and liver.

Nearly all the cultures from No. 402 contained a large spore-bearing bacillus which I have frequently found since then in cases of swine disease having hemorrhagic lesions. Cultures from the liver and blood contained also swine-plague bacteria which were isolated by rabbit inoculation.

It will be noticed that in the later cases, Nos. 403, 405, and 402, the large intestine was the seat of a peculiarly intense inflammation, accompanied by the exudation of circumscribed masses of fibrin easily lifted away from the mucosa, leaving a paler, slightly depressed spot showing no necrosis of tissue. In the rectum this exudate formed a continuous sheet also easily removable.

It might be claimed that the feeding of a dose of aloes or salts may have caused this peculiar intestinal inflammation. This claim, however, is effectually disposed of by case No. 407. This animal, one of the same herd, had not been taken from the neighboring farm. It was found dead February 22, and brought to the Station for examination. The notes of this case are reproduced because it must be considered the most trustworthy of all.

No. 407. Pig of medium size, white ; skin of abdomen, chest, neck, and back deeply reddened. Fat abundant, slightly reddened along the linea alba. Superficial inguinal glands slightly enlarged ; spleen dotted with elevated blood-red points.*

---

* These points are found in spleens of healthy swine.

Stomach and duodenum normal, the latter bile-stained. In ileum Peyer's patches are visible as groups of small, dark dots; no swelling. Mucosa of cæcum and upper colon of a dirty blackish color, probably pigmented. A few hæmatomata beneath mucosa. Besides the diffuse pigmentation the mucosa is sprinkled with isolated or confluent masses, about one-eighth to one-fourth inch in diameter, of a dirty grayish-yellow color, loosely adherent to the membrane. When pulled away a slightly depressed surface is exposed. Much of this mass can be easily removed by simply moving the scalpel over it. There are several ulcers in the cæcum with decided loss of substance. The patch of mucous glands at the base of the valve is also converted into an ulcerated mass. Lymphatic glands in abdomen slightly swollen and reddened. Kidneys deeply reddened to tips of papillæ.

On opening the thorax the lungs did not collapse, and a rather disagreeable odor was perceived. As in No. 405, the ventral and cephalic lobes of both lungs were consolidated. The hepatized regions were very hard to the touch, bright red, with yellowish points sprinkled in regularly. (See Plates III and IV.) The right lung was adherent to chest wall over the hepatized portion. A whitish, spongy membrane was interposed, about one-eighth to one-fourth inch thick, inclosing a small quantity of turbid liquid. On removing the lungs the membrane remained adherent to the pulmonary pleura, and was removed with difficulty. A portion of the diaphragm was also firmly attached. The left lung adhered firmly to the chest wall in two places, where it was consolidated. The costal pleura was deeply reddened, owing to the injection of a close net-work of minute vessels. Trachea and bronchi filled with whitish foam.

On section, the consolidated region is sharply but irregularly marked off from the normal tissue, very consistent and slightly elevated. The color varies from a bright red to a grayish red. In all, minute grayish points are present one-twelfth inch in diameter, about the same distance apart, and of a hazy outline. The smaller bronchi are filled with a purulent fluid. In the surrounding lobules in which the disease is more advanced the interlobular tissue is distended with a serous infiltration; the large vessels are filled with very consistent dark clots. Heart rather large, pericardium free; right auricle, ventricle, and large veins distended with clots; small white clot in left ventricle.

Microscopic examination of the lung tissue in cover-glass preparations shows the presence of numerous bacteria with the polar stain in recent lesions; in older ones they are rare. Other forms are present, but only in small numbers. The pleural exudate was made up of rounds cells, bound together by bundles of fibrin; it contains few bacteria.

In transverse sections of the large intestine, where a mass of exudate is still attached, the muscular and submucous layers are intact, if we except a slight cellular infiltration near the base of the crypts. The mucous layer, however, is considerably changed. The surface epithelium, including a portion of the crypts of Lieberkühn, is no longer distinguishable, but merges without demarcation into an exudate several millimetres thick, consisting of leucocytes imbedded in a mesh-work of fibrin, the whole refusing to stain.

Pure cultures of swine-plague bacteria in tubes of gelatine were obtained from the pleural exudate. In each needle track a large number of colonies developed. A piece of false membrane gave the same result. Cover-glass preparations from spleen and liver were negative. Two tubes of beef infusion into which bits of spleen had been dropped remained sterile. Two similar cultures from the liver contained each a large bacillus, evidently of post-mortem growth. The blood from the heart was also free from bacteria, for two tubes of gelatine, each inoculated six or seven times with blood, did not develop a single colony.

A rabbit inoculated in the ear with a bit of lung tissue died within 4 days. There was no swelling or reddening of the ear. Lungs deeply congested (hypostatic?). Immense numbers of swine-plague bacteria in blood, spleen, and liver. Cultures from blood and liver contained the same organisms. A mouse inoculated with a bit of lung tissue succumbed within 2 days. Bacteria very scarce in body. Pure cultures of swine-plague bacteria were, however, obtained from heart's blood.

What is of importance in this case and in No. 402 is the absence of hog-cholera bacilli from the internal organs where we would certainly expect to find them in this disease. If we refuse to consider the intestinal lesions as caused by swine-plague bacteria, we have the alternative of assuming the existence of bacteria which produce intestinal disease without penetrating into the internal organs proper.*

Five other cases in this investigation deserve special attention. Nos. 408, 409, and 410 of the same herd still remained on the farm. After the examination of 407 had shown the absence of hog cholera, and the presence in the diseased lungs of swine-plague germs, and the evidence thus far obtained pointed to a different disease caused by the latter bacteria, two pigs, Nos. 359 and 360, were taken from the Station to the farm and penned with the three mentioned to determine whether the disease is readily communicable. This was done February 28. March 5, No. 408 died. March 16 the remaining four were taken back to the Station and placed in an unused pen free from infection, so as to be under observation.

Of these four pigs No. 409 died March 20, No. 410, March 29. Of the exposed animals No. 359 died March 24, 24 days after the commencement of the exposure ; No. 360, April 6, 37 days thereafter. Of these four cases No. 360 was not examined. Of the remaining three, hog-cholera bacilli were detected in 409 by rabbit inoculation, but not in either 408 or 410, while swine-plague bacteria were found in 408 and 359. Looking over the original notes I am convinced that the number of cultures made from the spleens of 408, 410, and 359 were sufficient to enable us to exclude the presence of hog cholera from these cases. No. 409 is therefore the only animal from the adjoining farm which was examined in which hog-cholera bacilli were detected. It is likewise strange that in the spleen of No. 408 a bacillus should appear resembling hog-cholera bacillus in many respects, but not identical with it, and producing only suppuration in rabbits.

The cases which came under observation subsequently, and in which hog-cholera bacilli were readily demonstrated in the spleen, were Station pigs which had been exposed some time ago to the infection of hog cholera. These cases are chiefly valuable in pointing out that the swine-plague bacteria are transmitted from animal to animal and associated with lung disease. They may be thrown out altogether, since the source of the hog-cholera bacilli is traceable.

No. 372. Fed viscera of hog cholera case December 24, 1886, in infected pen.

---

* By "internal organs" I mean those organs which do not come in contact with the air or food, either directly or indirectly. Thus the entire respiratory and digestive tract may be regarded as external, so far as bacteria are concerned.

No. 378. Fed viscera of hog cholera case November 18, 1886, in infected pen. Fed viscera of 406 February 2, in infected pen.

No. 392. Inoculated with swine plague bacteria (Iowa) January 25; placed in the infected pen March 28.

No. 397. Fed viscera of 378 in pen 7 March 24.

In 372, 392, 397, both hog cholera and swine plague bacteria were found. Their presence in the old infected pen or their contact with pigs kept there will account for this double infection.

The facts brought out by this investigation corroborated those already brought out in the investigations of the preceding year. They showed the existence of pneumonia and pleuritis, together with intestinal disease in most of the animals examined, associated with bacteria readily distinguished from hog-cholera bacteria.

They also demonstrated the transmissibility of the pneumonia to other pigs, and in these pigs the same bacteria were found.

If the lesions of the large intestine, as observed in the early cases, were due to hog-cholera bacilli, why were these bacilli not found excepting in one late case in which there may have been an accidental infection of the rabbit inoculated? Why were hog-cholera bacilli readily detected in the later cases kept in the infected pen?

While these investigations do not prove that the swine-plague bacteria were the cause of the intestinal lesions, they also do not seem to show any relationship between these lesions and hog-cholera bacilli. This question of the relation between intestinal lesions and swine-plague bacteria will be discussed farther on.*

V.

In September, 1888, an outbreak of swine disease near Baltimore, Md., came to our notice. Three pigs from one herd and one from another herd were examined. The following synopsis of the cases may be of interest, the full account being given elsewhere.†

Pig No. 1. Broncho-pneumonia; exudate on mucosa of rectum and lower colon; swine-plague bacteria detected in lungs and rectum; hog-cholera bacilli in spleen.

Pig No. 2: Three-quarters of both lungs hepatized; ulcers in ileum and colon; swine-plague bacteria in lungs; hog-cholera bacteria in spleen.

Pig No. 3: Extensive pneumonia and pleurisy; ulceration of large intestine; swine-plague bacteria in lungs; hog cholera bacteria in spleen.

Pig No. 4: Slight atelectasis of lungs; ulcers in large intestine; swine-plague bacteria in large intestine; hog-cholera bacteria in spleen.

Pig No. 3 was taken to the Experiment Station, where it became the starting point of a mixed outbreak of hog cholera and pneumonia.

---

* P. 102.
† Report of the Bureau, etc., 1887-'88, p. 121.

VI.

During November, 1888, the writer was directed to make some investigations in Iowa,* where swine diseases at that time were prevailing to a considerable extent. In the vicinity of Mason City pigs from three herds, some distance apart, were examined. The lesions observed were both intestinal and pulmonary. The intestinal lesions, though varying considerably in appearance and intensity, did not differ, on the whole, from those observed in hog cholera. The lung lesions varied considerably in character and extent, from a slight collapse in a single lobe to almost total hepatization, accompanied by exudative pleuritis.

The investigations were limited to post-mortem examination and cultures from the spleen. From each spleen small bits of tissue were removed to two tubes of agar. In addition, portions of diseased lung tissue and ulcerated mucous membrane from the large intestine were cut out and transferred to sterile test tubes plugged with cotton wool for inoculation into rabbits. Ten pigs were examined in this manner.

In but one case did the spleen pulp show any bacteria under the microscope, and in this case they were streptococci. In but one agar tube of the spleen series did anything develop. This was a motile bacillus, resembling the hog-cholera bacillus in form, but differing in its growth on agar, in bouillon, and in gelatine, this growth being in all cases more vigorous. Of two mice and one rabbit inoculated, one mouse died in 5 days. The lesions were indefinite and did not point to hog cholera. Moreover, the other mouse and the rabbit remained unaffected.

With particles of diseased lung tissue and mucous membrane from some of these cases a considerable number of rabbits and some mice were inoculated. Of those that died some contained no bacteria of any description. Those inoculated from three cases out of ten died of swine plague since these bacteria were found. A few survived the inoculation.

The nature of this disease was not, therefore, cleared up by these investigations, since the results were not uniform. While in hog cholera the bacilli are present in the spleen, and readily obtained therefrom by cultivation, they were not present in the spleens of these ten cases. As already stated, swine-plague bacteria were obtained from three cases. They were quite virulent, as the inoculations upon pigs show. Thus one pig, which received 9 cubic centimetres of a bouillon culture into the right lung through the chest wall, died within 20 hours of septicæmia, the injected bacteria being present in the spleen in considerable numbers. Another pig, which received a subcutaneous injection at the same time, remained well. Somewhat later two pigs received into the right lung 1½ cubic centimetres and 3 cubic centimetres of a bouillon

---

* Report of the Bureau of Animal Industry for 1887-'88, pp. 135-145, where a detailed account of this work is given.

culture. The one which had received the smaller dose was sick for a time, but recovered. It was killed 1½ months after the inoculation, and both lungs were found everywhere adherent to the chest wall. In the pericardial sac a considerable quantity of pus had collected, in which the injected bacteria were still present, as determined by cultures.

The animal which had received the larger dose became very ill and was killed 5 days after inoculation. At the autopsy were detected partial hepatization of the right lung, with extensive exudative pleuritis and pericarditis.

The two following cases, which have not yet been published, demonstrate very strikingly the virulence of these bacteria. The growth on a number of agar cultures about 9 days old was scraped off and transferred to sterile bouillon until a turbid suspension was obtained. This suspension was prepared because these bacteria multiply very feebly in bouillon, and when the latter is used as the injecting fluid very few bacteria are actually introduced into the body. This turbid suspension was used to inoculate 2 pigs. No. 120 received into the abdomen 2 cubic centimetres; No. 143 into the right lung through the chest wall but 1 cubic centimetre. The inoculation was made March 11, 1889, over 3 months after the bacteria had been obtained from the diseased swine.

No. 120. Essex female, 5 months old. Into the right lung through the chest wall, 2 cubic centimetres of above suspension injected with hypodermic syringe, 6 p. m. March 11. Found dead early next morning.

Subcutaneous inguinal glands partly hemorrhagic. From the cut subcutaneous vessels of right side liquid blood oozes out. In abdomen the solitary follicles of large intestine appear as bright red circular spots three-sixteenths inch diameter, as seen from serosa. Spleen slightly engorged.

Large quantity of blood-stained serum in right pleural sac. The needle had punctured the convex surface of principal lobe, where there was some infiltration of blood. A thin layer of fibrin on convex surface of the small anterior lobes and on pericardium. A few collapsed lobules along free border of these lobes.

In stomach, the mucosa of fundus is deeply reddened over an area of 3 to 4 square inches; in this area two hemorrhagic spots. The upper half of duodenum with mucosa intensely reddened. Contents somewhat blood-stained. All Peyer's patches in the small intestine from duodenum to ileo-cæcal valve are intensely reddened, the follicles appearing as blood-red points. On some patches, hemorrhages on the surface. Considerable number of ascarides in small intestine. In large intestine the lymphatic patch near valve likewise reddened. Mesenteric and mesocolic glands hyperæmic.

Kidneys very much congested. The glomeruli appear as minute blood-red points. Small quantity of urine in bladder loaded with albumen. Blood fails to clot. In the spleen large numbers of swine plague bacteria.

No. 143. Essex male, 5 months old. Injected into abdomen 1 cubic centimetre of turbid suspension of swine plague bacteria, March 18. Animal dies 40 hours after inoculation.

Inguinal glands slightly swollen and hyperæmic. In abdominal cavity peritoneum pale pink, all minute vessels injected. Shreds of a viscid grayish exudate attached to abdominal walls, contiguous coils of intestines, and stomach. Considerable opaque reddish serum present. Vessels of diaphragm injected; some ecchymoses observed. The shreddy exudate also present. The mesentery œdematous, especially where attached to intestines. The walls of a portion of the lower small intestine very much swollen, serosa dark red. On the mucosa which is congested a yellowish, pasty exudate loosely rests, occupying the side to which mesentery is attached. This exudate is made up of leucocytes imbedded in strands of fibrin. Peyer's patches along this region and down to ileo-cœcal valve are deeply congested and swollen so as to appear boat-shaped. (The bacteria had evidently traveled along mesentery and invaded the walls of the small intestine.)

Large intestine distended with dry feces. Mucosa of cœcum and colon more or less congested, the congestion limited mainly to summits of folds. The lymphatic patch near valve very hyperæmic and swollen.

In each pleural sac about 50 cubic centimetres of reddish serum. A thin membranous exudate covering the dependent half of both lungs easily scraped away as a pale yellowish pulpy mass. The remainder of pleura opaque, a barely visible exudate covering it. The lungs are hyperæmic, the free border of the ventral and cephalic lobes collapsed. Bronchial glands swollen and hyperæmic.

Pericardium thickened, clouded; vessels injected. A barely visible exudate on it. Vessels of heart surface very much distended. Petechiæ on left auricle and under endocardium of left ventricle near semi-lunar valves. Right heart distended with a dark, soft coagulum.

In the peritoneal exudate, which consists chiefly of fibrin and a few leucocytes, immense numbers of the injected bacteria are present, showing after staining the polar arrangement of protoplasm very distinctly. The pleural exudate composed of the same elements. In some leucocytes up to twenty bacteria. In the spleen and blood from the heart large numbers of swine-plague bacteria. Agar cultures from the spleen, pleural, and peritoneal cavities contain only the injected bacteria. A bouillon culture from the spleen likewise pure.

The problem of swine diseases as it stood after the completion of these investigations up to 1889 may be stated briefly as follows:

Since 1885 a well-characterized bacillus has been encountered as the cause of an infectious disease termed hog cholera, which is chiefly localized in the large intestine. Since 1886 our attention has been directed to lung disease in swine with which a bacterium is associated, which, when inoculated into swine, proves to be very virulent and may give rise to pneumonia when the bacteria are injected into the lungs. This is sufficient to demonstrate the existence of a disease differing from hog cholera, which has been called swine plague because an identical disease of swine in Germany, first described in 1885, was called *Schweineseuche*. This lung disease was shown to be communicable. (IV.) *

In many of the outbreaks examined the changes found in the intestines could not be distinguished from hog cholera (except perhaps in IV). In some hog-cholera bacilli were actually detected, in others (I, VI) they could not be found. One of the problems, therefore, still before us, and a very important one, was to determine whether all outbreaks of swine plague in which intestinal lesions closely resembling those of hog cholera are present are mixed outbreaks of swine plague and hog cholera, or are simply swine plague.

Much light has been thrown upon this subject by the investigations of three outbreaks given in detail in the following pages. The first is a mixed outbreak in which, however, the lung lesions are so very well marked and the swine-plague bacteria associated with these lesions so virulent that there can be little doubt that the hog cholera disease was really secondary to the swine plague. The second outbreak is simple, uncomplicated swine plague. In the third outbreak very virulent swine-plague bacteria, and, in one case, very attenuated hog-cholera bacilli were found.

VII.

An outbreak of swine disease appeared among the pigs belonging to an abattoir adjoining the Experiment Station about the first week in October, 1889. The disease came to our notice a week after the first

* These numerals refer to the different outbreaks as numbered in this report.

animals died, and in attempting to trace the causes which led to it we obtained the following information:

The pigs were purchased in the markets of Washington City the latter part of September. They were thirty-five in number, in two, possibly three, lots. One lot came in two crates. It could not be determined whether it was made up of pigs from one or two sources. They began to die, as stated, about a week after their arrival. We examined in all thirteen animals, the first on October 12, the last on October 28. A few days later the last of these thirty-five pigs succumbed to the disease. It lasted, therefore, about 1 month. A few large swine which were on the place when these animals arrived did not take the infection.

In the following pages a detailed statement of the pathological and the bacteriological examination is given. Those who are not specially interested in these notes will find a summary of the ascertained facts following them.

No. 1. October 12. Male pig, weighing about 25 pounds, died last night. On inner aspect of right thigh an area, about one-half inch in diameter, of extravasation, extending down into true skin. No ulcers in the mouth. Spleen not enlarged.

Contents of stomach slight in amount, consisting chiefly of sand and a turbid liquid; walls contracted, throwing mucosa into large folds. The greater part of mucosa intensely congested. In the fundus a large ulcer 1½ inches across, covered by a slough one-quarter inch thick. The subjacent wall nearly one-half inch thick, on section deeply reddened throughout, due to a sanguinolent, œdematous infiltration of the wall. Adjacent to this a smaller inflammatory thickening covered with a thin pultaceous slough. The mucosa of duodenum uniformly and deeply pigmented. In the ileum the mucosa is swollen, Peyer's patches reddened.

Mucosa of cæcum of a bluish-gray color, dotted with a small number of ulcers not much larger than pins' heads, covered with yellowish sloughs. Similar follicular ulcers on Peyer's patch over valve. Upper colon contains much earth, adhering rather closely to mucosa. The latter quite deeply pigmented, the pigmentation intensified in lower colon. In upper colon a small number of flattish yellowish-white sloughs from one-sixteenth to one-eighth inch in diameter.

In thorax, the pleura covering the diaphragm and ribs is overlaid by a pale pinkish membranous exudate, easily rubbed off and especially abundant on the right side.

The various lobes of both lungs firmly glued together by exudate. The ventral lobes, the major portion of cephalic lobes, and a small portion of the principal lobes adjacent to the ventrals are hepatized. The hepatized areas are covered by false membranes varying in thickness and easily peeled off. Through the hepatized lobes are disseminated necrotic masses of a greenish color varying in size from mere specks to peas. They contain large numbers of swine-plague bacteria, which show a polar stain very clearly. The tissue around the terminal portion of both bronchi in the principal lobes is hepatized and contains necrotic foci. Lung worms abundant in the left bronchus. Both bronchi contain small quantities of reddish foam. Pericardium thickened, opaque; vessels injected; the ventral surface is covered by a membranous exudate and it is adherent to the heart surface by means of a similar exudate. Cover-glass preparations from the pleural and pericardial exudate, from various regions of hepatized lung tissue, contain large numbers of swine-plague bacteria.

A rabbit was inoculated October 12, by placing a bit of lung tissue under the skin of abdomen. Dead next morning. Slight ecchymosis in the subcutis at the point of inoculation. The blood and spleen contain immense numbers of swine-plague bacteria, showing well the polar arrangement of the protoplasm when stained. Cultures confirmatory, agar and bouillon being used chiefly.

A rabbit inoculated at the same time and in the same way with a bit of spleen tissue. Rabbit dead next morning. Swine plague bacteria fewer than in preceding case. An agar culture from the blood contains only these germs. From the spleen two agar, one bouillon, and one gelatine tube culture prepared. On the second day (Sunday intervening) a moderate number of isolated colonies on the agar surfaces: the condensation water turbid with flaky deposits. In the bouillon tube are a large number of minute suspended flakes, the liquid itself not clouded. In these cultures only swine-plague bacteria. The gelatine tube remained sterile.

From the pleural exudate an agar and a bouillon tube were inoculated at the autopsy. On the second day a number of colonies similar to those on the spleen agar culture were found to be made up of swine-plague bacteria. The bouillon culture faintly and uniformly clouded. Only swine-plague germs present. From the epicardial exudate two similar cultures were prepared with the same result, with the exception that in the bouillon culture large cocci were also present.

With a bit of lung tissue taken from more recently diseased regions two gelatine rolls and two agar plates were made. In the second gelatine roll about 100 minute colonies present after 3 or 4 days. These were examined and bouillon tubes inoculated from four different colonies at different times. In all the swine-plague germs only appeared. The first agar plate developed a very large number of colonies, the second only six or seven, made up only of swine-plague germs.

No. 2. Female; died last night; weighs about 35 pounds. Lymphatics in the groin barely enlarged, pale. Spleen quite large, softened, dark red in color.

Digestive tract: Stomach partly filled with chewed leaves, and straw adhering slightly to mucosa. Fundus faintly reddened. The pyloric portion bile-stained. Duodenum similarly stained. An ascaris in the bile duct, projecting 1½ inches out into duodenum. Arborescent injection of mucosa of the latter. Considerable quantity of turbid liquid in large bowels. Mucosa not pigmented. Peyer's patch at base of valve thickened and pigmented. Ulcers moderately abundant in cæcum and upper colon. They appear as little cup-shaped depressions, one-sixteenth to one-eighth inch in diameter, lined with a pale yellowish necrotic layer, the center in some filled with a black slough and the margin elevated. Mesenteric, mesocolic glands, and those of lesser omentum, slightly swollen; cortical layer congested. Interlobular markings of liver prominent, broadened. The section appears mottled, some lobules being much more congested than others. The empty gall-bladder contains a partly disintegrated ascaris, which extends through bile duct into duodenum.

Thorax: The costal pleura of both sides highly injected and covered with a thin, whitish membranous exudate, gluing the lungs to the chest wall. Of the lungs all but a small strip along the dorsal border of the principal lobes is solid, three or four times the size of the normal lung when collapsed. The various lobes are glued to each other, forming a single mass, in which the individual lobes are not recognizable, and which is in turn attached to the chest wall, the diaphragm, and the pericardium by pleuritic exudate. This is most abundant, and forms membranes on the most dependent portions of the lungs, easily pulled away from the subjacent structures. The various adhesions torn without difficulty.

The disease seems farthest advanced in the cephalic and ventral lobes, in which are imbedded a large number of closely set masses of dead tissue, from one-sixteenth to one-half inch in diameter. These are firm, yellowish masses, sharply defined in outline, and imbedded in dark red hepatized tissue, which is mottled with paler specks, representing the ultimate lobules distended with cell masses.

The trachea and bronchi contain whitish foam, mixed with large quantities of thick, purulent secretion. In the bronchi of the principal lobes are many lung worms. The bronchial glands are hyperæmic and œdematous, and contain a small number of necrotic foci.

In the spleen a few minute oval bacteria detected. With a platinum needle the surface of an agar tube and a bouillon tube inoculated. On following day the former showed a considerable number of round, grayish colonies, 1½ to 3 millimetres (one-sixteenth to one-eighth inch) in diameter; condensation water clouded. In this tube only swine-plague germs detected. The bouillon culture contains a considerable number of minute granules, representing clumps of swine-plague bacteria.

An agar culture from the liver grew like the spleen culture and contained only swine-plague germs.

At the autopsy an agar and a bouillon tube inoculated from the right pleura. In the former only the condensation water became turbid and contains swine-plague germs exclusively. The bouillon tube remains sterile; similarly a bouillon tube from pericardial exudate remains sterile. (It is highly probable that the flocculent growth of the swine-plague germs in bouillon and the sterility of these tubes was occasioned by an unsuitable condition of the bouillon.)

From a bit of lung tissue, the pleural surface of which had been thoroughly scorched, gelatine roll cultures were prepared. After a number of days the first roll showed a large number of colonies as mere points, the second roll about 100 colonies, somewhat larger. From both tubes bouillon was inoculated from individual colonies. These contained after development only swine-plague bacteria.

At the same time the scorched lung tissue was pricked with platinum needle and an agar tube inoculated. On the surface a considerable number of similar colonies appeared on the following day. Some of these examined were swine-plague germs. In the condensation water a few chains of a short, motionless rod appeared among the great bulk of swine-plague germs.

At the same time a rabbit was inoculated from the lung tissue. It was dead next morning. Enormous numbers of polar stained swine-plague germs found in stained preparation of spleen pulp. An agar culture confirms the microscopic examination.

No. 3. Died last night. Female; weighs about 25 pounds; in very poor condition. Spleen considerably engorged; contains a small number of bacteria; nature to be determined by cultivation.

Digestive tract: One ulcer on the gums of lower jaw. Stomach contents like those of No. 2. In the fundus an area of mucosa, about 4 inches in diameter, is nearly black from extravasation, and thickened. A zone several inches wide outside of this is deeply reddened. The mucosa of duodenum and ileum more or less discolored the vessels, showing arborescent injection. In the cæcum, the mucosa is of a bluish-gray color, and dotted with a dozen sloughs, one-eighth inch diameter, of a yellow color; the tissue at the margin of the slough thickened, elevated. The Peyer's patch near valve deeply congested and sprinkled with enlarged grayish follicles. Mucosa of colon more or less discolored and hyperæmic in patches; only three small ulcers found. The liver appears normal, the bile in bladder very thick, flaky. Punctiform ecchymoses on surface of kidneys.

In the lungs the major portion of both ventral lobes is solidified, the tips being emphysematous. The hepatized portions are bright red in color, with faint yellowish mottling.

Bacteriological examination: Cover-glass preparations of the hepatized lung tissue show no bacteria. From a bit of tissue two gelatine roll cultures prepared and a rabbit inoculated. Roll A subsequently contained a large number of apparently identical colonies. Roll B, about fifty of the same colonies and one chromogene. Careful examination of these revealed *hog-cholera* bacilli. The inoculated rabbit died within 36 hours. At the place of inoculation the bit of lung tissue was surrounded by a small area of purulent infiltration and dilated blood-vessels. In the blood and spleen a very large number of swine-plague bacteria. A bouillon culture from the blood faintly clouded on the following day, and holding in suspension barely visible granules made up of clumps of swine-plague bacteria. No motile bacteria detected. The spleen of the pig contained a few bacteria, character not determinable. With a bit of pulp a gelatine roll A and agar plate B prepared, also a bouillon culture. The latter was uniformly clouded on the following day ; a few gas bubbles on the surface. Only motile hog-cholera bacteria present. The presence of hog-cholera bacilli in the spleen was furthermore demonstrated by the roll and plate culture. The agar plate grew, of course, most rapidly, being in the thermostat. On the second day a considerable number of colonies appeared, apparently the same. Examination of some showed only hog-cholera bacilli. A bouillon tube inoculated from one of them was clouded on the following day with motile hog-cholera bacilli.

One-fifth cubic centimetre of the original bouillon culture was injected subcutaneously into a rabbit. On the fourth day temperature 106.1° F. Found dead on the seventh day with the usual lesions of hog cholera, such as engorged spleen, necrotic foci in the liver, reddened Peyer's patches, hemorrhagic points on the lower colon and rectum. Spleen and liver contain hog-cholera bacilli in moderate numbers. From spleen an agar and a bouillon culture prepared by piercing spleen with a fine platinum needle and inoculating culture media therewith. In both only hog-cholera bacilli appeared. The original spleen culture contained no swine-plague germs, otherwise the rabbit would have succumbed within 1 or 2 days.

In the gelatine roll appeared after a few days a very large number of the same colonies, evidently all hog cholera.

No. 4. October 14. Small male pig, weighing about 30 pounds ; died last night. Skin of abdomen slightly reddened. Subcutaneous inguinal glands enlarged, œdematous, mottled red and pale.

Mouth free from ulcers. Contents of stomach stained with bile. Mucosa of fundus over an area 4 inches square deeply congested, swollen. Slight injection of minute vessels in duodenum ; remainder of small intestine normal. Large intestine contains a turbid liquid and a substance resembling coal ashes. The mucosa of cæcum studded with superficial, yellowish sloughs, about the size of pins' heads or a little larger. In the colon, besides these small yellowish sloughs there are three or four large areas over which the mucous membrane is entirely necrosed superficially. These areas are about 6 inches long. The minute sloughs limited chiefly to the upper half of the colon.

Liver rather firm in texture, the acini projecting slightly above interlobular tissue on the surface. Glands near the portal fissure deeply congested throughout.

Dilatation of pelvis of kidneys encroaching on medullary portion to a large extent. Both ureters very much distended ; walls from one-half to three-quarters inch thick.

Lungs collapse when removed from thorax ; no pleuritis. In all the lobes there are masses of collapsed and hepatized tissue varying in diameter from one-half to several inches. They are situated chiefly near the roots of the lobes. The small median lobe in part hepatized, bright red with yellowish mottling. In the terminal bronchi are plugs of lung worms imbedded in thick muco-pus. The tissue around the plugs in the left principal lobe hepatized.

Bacteriological examination : With bits of diseased lung tissue two gelatine rolls were prepared and a rabbit inoculated. The rolls both contained on the second day an immense number of colonies of micrococci, evidently a contamination of the gelatine. The inoculated rabbit died in 4 days with enlarged spleen and patches of necrosis in the liver in the form of a net work. An agar culture from the spleen and a bouillon culture from heart's blood both subsequently contained only hog-cholera bacilli. No swine-plague germs detected, although the rather premature death of the rabbit led me to suspect their presence. Some weeks later a rabbit was inoculated subcutaneously in the ear with a platinum loop of the blood culture of the preceding rabbit. This rabbit died in 5 days with hog-cholera lesions.

With a bit of spleen pulp of the pig a gelatine roll A and an agar plate B were prepared ; also a bouillon tube inoculated with platinum needle after pricking the spleen. This developed into a pure culture of hog-cholera bacilli. In the gelatine roll appeared numerous colonies of hog-cholera bacilli. The agar plate was partially overgrown ; the few isolated colonies were made up of hog-cholera bacilli.

From the liver a gelatine roll, an agar and a bouillon tube inoculated, each with minute particles of tissue. In the roll numerous colonies appeared, evidently all hog cholera. The two tube cultures likewise contained only hog-cholera bacilli.

No. 5. October 15. Small male pig, weighing about 25 pounds, died last night.

Skin about urinary meatus reddened. Subcutaneous inguinal glands enlarged, firm, juicy, faintly mottled with red.

Stomach contracted. Contents same as in preceding cases. Mucosa in large folds, summits of which somewhat reddened. Slight discoloration of mucosa of lower ileum. Contents of large intestine firm, in form of lumps. In the cæcum about ten ulcers, from one-eighth to one-half inch in diameter, the largest button-shapped, flattish, a firm, yellowish, necrotic base extending into subjacent muscular coat. Complete necrosis of Peyer's patch at valve. In the colon a small number of large and small button ulcers. The solitary follicles projecting as large as split peas. When squeezed a white soft mass exudes from a central depression.

Liver rather firm in texture. Gall bladder contains semi-liquid, flaky bile.

Spleen slightly enlarged. Kidneys on section very pale.

Both lungs œdematous. Hepatization involves the free tip of both small cephalic lobes of the right lung, the major portion of the ventral lobe of the left lung. In both principal lobes are a moderate number of small hepatized foci. In both bronchi a large quantity of very viscid muco pus, which extends down into the branches of the principal lobes. These latter and the terminal portion of the principal bronchi contain plugs of lung worms. Bronchial glands enlarged, firm, pale.

Bacteriological examination: In cover-glass preparations from the spleen a very few bacteria resembling somewhat hog-cholera bacilli seen. An agar culture inoculated with platinum wire remained sterile. On the following day spleen taken from refrigerator and a second agar tube inoculated with particle of pulp. In this tube a colony of greenish fluorescent bacilli appeared.

From hepatized lung tissue two gelatine rolls were prepared. The first one was spoiled by a few liquefying colonies: the remaining small colonies were inaccessible. The second roll remained sterile.

A rabbit inoculated subcutaneously with a particle of lung tissue died in 24 hours. At the place of inoculation considerable purulent thickening of skin with ecchymosis of the subcutis. Numerous very small coccidia cysts in liver. Spleen enlarged, congested. (Bacteriological notes of this rabbit mislaid.)

No. 6. October 15. Small female in very poor condition; died last night; more or less decomposition. Ventral aspect of body reddened. Subcutaneous inguinal glands enlarged, firm, and very hyperæmic.

Stomach much contracted, empty. Mucosa of fundus beset with punctiform hemorrhages. Small intestine not marked by changes; one ulcer in lower ileum. Contents of cæcum and colon of a somewhat pasty consistency mixed with coarse sand. In the cæcum about six old, flattish ulcers, from one-fourth to one-half inch in diameter, also a patch of easily removable, membranous exudate several inches square. In the colon near the valve several ulcers and a similar patch of exudate.

In the lungs the hepatization, though extensive, is more or less scattered in foci as follows:

A few hepatized foci in left cephalic, one large hepatized mass in ventral, and four wedge-shaped masses in left principal lobe. In the right lung, fully two-thirds of the median and the whole of the ventral lobe solidified. The latter lightly glued to the neighboring lobes. Disseminated through it are very many necrotic foci about one-eighth inch in diameter. In the right principal lobe several hepatized masses. In both bronchi a large number of adult lung worms. Pericardium thickened; vessels injected. Spleen not enlarged or congested.

Bacteriological examination: Spleen found more or less decomposed. No germs in cover-glass preparation. In an agar culture therefrom numerous isolated and confluent hog cholera colonies appeared on following day.

A rabbit inoculated subcutaneously with a bit of lung tissue (the particular region not indicated in the notes) was found dead on the eighth day. Extending from the point of inoculation over the entire abdomen and portion of thorax the subcutis is infiltrated with a yellowish, pasty mass of pus, stained with blood. The superficial layer of muscles discolored, ecchymotic. Spleen slightly enlarged. No peritonitis. Cultures from the spleen remained sterile. An agar plate from the subcutaneous exudate spoiled by condensation water. An agar tube from the same source remained sterile.

No. 7. October 15. Small male pig, of about 35 pounds, died yesterday afternoon late and put in refrigerator until next morning. Animal in fairly good condition. The lesions in this case were briefly as follows:

Digestive tract: Fundus of stomach faintly reddened. Small intestine not affected. Contents of large intestine liquid, of black color, holding much earth in suspension. One ulcer on the thickened Peyer's patch near the valve. Follicles swollen. No other changes.

Lungs: General pleuritis indicated on the most dependent portions of the left lung by a thin, membranous exudate, elsewhere by roughening. The various lobes lightly glued together. Hepatization limited to the tip of the left ventral and a small area in the left principal lobe, containing numerous necrotic foci. In the right principal lobe two wedge-shaped, hepatized masses, in which are many minute oval germs, resembling swine plague. Generalized bronchitis indicated by much curdy muco-pus in the bronchi and branches. No lung worms detected.

In the liver, the center of many lobules in all the lobes of a brick-red color, caused by fatty degeneration of this portion of the lobule. Bile very thick, dark-colored.

Spleen very much enlarged, softened, dark colored.

Bacteriological examination: From this animal only the spleen and the pleural exudate received any attention. An agar tube inoculated from the spleen simply with a platinum wire was sterile on the following day, although bacteria had been detected in the spleen pulp in cover-glass preparations. A second tube inoculated with a particle of pulp from the spleen, kept meanwhile in the refrigerator, developed a large number of isolated and confluent colonies. So far as could be determined these were all swine-plague bacteria. No motile bacteria present.

An agar culture made at the autopsy from the pleural exudate proved to be a pure culture of swine-plague bacteria.

No. 8. October 17. Male pig, weighing about 35 pounds, died last night. Subcutaneous inguinal glands enlarged, on section dotted with minute petechiæ.

Stomach normal. Duodenum bile-stained. In lower ileum much liquid, containing fine earth and sand. In cæcum minute vessels of mucosa injected. Peyer's patch at valve somewhat swollen and discolored. In the upper colon several small patches of thin exudate, about one-fourth inch in diameter; the subjacent epithelium appeared necrosed. In both kidneys infarcts of a yellowish, homogeneous appearance, surrounded by a hyperæmic zone, and visible from the surface of kidney, two in the left and twelve in the right.

Spleen somewhat enlarged, firm. Liver very pale, pulp rather soft. On the left lobe several paler spots, not elevated, about one-fourth inch in diameter. Bile thick, flaky.

Lungs everywhere adherent to surrounding structures by means of a membranous exudate, grayish, elastic, coming away in patches and shreds when pulled. This membrane especially thick between lungs and diaphragm. The various lobes matted together and adherent to pericardium. Lungs do not collapse.

Left ventral and cephalic lobes completely hepatized. Throughout both are disseminated a large number of greenish-white, homogeneous, firm masses from one-fourth to one-half inch in diameter. Almost entire right lung hepatized. In the cephalic lobe, ventral lobe, and cephalic portion of principal lobe the necrotic masses are very numerous, one near tip of ventral lobe quite large. The hepatization of the principal lobe is of the dark-red variety.

Large quantities of lung worms in left bronchus and branches contained in the caudal third of principal lobe. Right bronchus not opened. Pericardium thickened, opaque. Heart surface covered with a thin, transparent pseudo-membrane. Left auricle hemorrhagic. A clot distending the right heart and forming of it a complete cast. Center of clot pale.

Bacteriological examination : From spleen two agar cultures made. One remains sterile. In the other on second day a faint growth starting from condensation water upwards. The latter contains clumps of swine-plague bacteria and large bacilli with terminal spore. Several gas bubbles in agar.

From the pleural exudate of right lung an agar and a bouillon culture prepared at autopsy. On following day a large number of punctiform colonies on agar surface. Minute flakes in condensation water. In bouillon minute granules, representing clumps of minute oval cocci. In both tubes only swine-plague bacteria.

From the more recently hepatized regions of lungs which contain large numbers of swine-plague bacteria and some large bacilli, two gelatine rolls and two agar plates prepared.

In the gelatine roll A two forms of colonies appeared, one with disk homogeneous, the other with a distinct peripheral zone. In roll B only one kind appeared. A number of bouillon tubes were inoculated from colonies in roll B which turned out to be streptococci. From roll A bits of gelatine were removed to bouillon with warmed platinum needle, some cultures remained sterile while others contained streptococci.

On the agar plate A, large numbers of apparently identical colonies appeared. On plate B, a moderate number developed. Of these a few examined were made up of swine-plague bacteria.

A large rabbit inoculated with a particle of lung tissue died within 20 hours. No internal changes, excepting a probably preëxisting fatty condition of the liver. Few swine-plague bacteria detected with the microscope. An agar and a bouillon culture from heart's blood contained only swine-plague bacteria on second day.

No. 9. October 18. Female pig; died last night ; weighs about 30 pounds. Subcutaneous glands of groin enlarged, firm, cortex hyperæmic.

One small necrotic patch on mucous surface of lower lip. Stomach with fundus pale, pyloric region bile-stained. Around the cardiac expansion are about thirty yellowish-white, confluent, and isolated ulcers from one-eighth to one-fourth inch in diameter. They are raised above the level of the mucosa, flat on top. Projecting slough soft, friable; base of ulcer very firm, extending into muscular coat. Mucous membrane of duodenum bile stained; arborescent injection of minute vessels. Mucosa of ileum more or less discolored and inflamed. Large intestine with walls very much infiltrated and mucosa extensively destroyed by necrotic changes. In the cæcum, a large patch of ulceration having a peculiar gnawed appearance, surrounding islands of intact mucosa. On section the mucosa is found converted into a yellowish-white, firm mass.

In the colon patches like these are interspersed with a large number of isolated circular ulcers with blackish, depressed surface and a subjacent yellowish, firm, thick base extending into the muscular coat. Besides these, there are a considerable number of ulcers with an elevated, soft, necrotic top, easily scraped away. In lower colon, large patches of destroyed mucosa.

Spleen very large, friable, pulp very dark. Liver tissue apparently unaffected. Bile rather thick. Kidneys with parenchymatous degeneration of cortex.

Lungs as whole much affected. In the right lung all but the dorsal third of the ventral lobe solid, enlarged, bright red, with large and small masses of a pale-greenish color disseminated through it. A portion of the cephalic lobe solid, in the same condition. One-half of the median lobe similarly diseased. The principal lobe glued to ventral lobe by a thin pseudo-membrane. About one-third along the ventral aspect solidified in masses from one-half to three-fourths inch in diameter, these masses extending from diaphragmatic to convex surface of the lobe. The interlobular tissue around these hepatized masses distended with serum. The various lobes of the left lung are in the same condition as the corresponding ones of the right lung. Pleuritis only over the hepatized regions in the form of a delicate network.

The trachea and bronchi contain much frothy muco-pus. In the left terminal bronchus, surrounded by hepatized tissue, are masses of lung worms completely occluding it.

Bronchial glands large, pale, containing a variable number of small necrotic foci.

Bacteriological examination: From the right pleural sac a bouillon and an agar tube were inoculated with platinum loop. In both only swine-plague bacteria subsequently appeared.

From a more recently hepatized region of the lungs two agar plates were prepared with a loop of the serum, in which microscopic examination had shown an immense number of swine-plague bacteria. Plate A contained, after several days, a small number of miscellaneous colonies and a very large number of barely visible colonies, which proved to be made up of swine-plague bacteria. On Plate B no development.

From the more advanced disease two gelatine rolls were prepared from a particle of necrotic tissue. Roll A liquefied. Roll B, after a week's time, contained about twenty-five small colonies of the same character, resembling hog cholera, and one large colony. The former transferred to agar, and bouillon tubes proved to be not hog cholera but streptococci.

A rabbit inoculated with a bit of diseased lung tissue died within 20 hours. Organs not changed. At point of inoculation considerable purulent infiltration. In blood and spleen a very large number of swine-plague bacteria, showing in stained preparations the polar stain very well. Cultures in bouillon and on agar confirmatory.

From spleen a particle of pulp used to prepare one bouillon culture and an agar plate. In the bouillon the motile hog-cholera bacilli appeared among clumps of swine-plague germs. On the plate after several days a moderate number of colonies of one size and a very large number of colonies of a very small size appeared. Some of the former examined are hog-cholera bacilli, the latter swine-plague bacteria.

From particles of liver tissue the same cultures were made, and the same mixture of hog-cholera and swine-plague bacteria found.

Sections from more recent pneumonic infiltration prepared from material hardened in alcohol show a considerable amount of fibrin in the alveoli, in whose meshes are a few leucocytes and very many of the minute swine-plague bacteria. The peribronchial and interlobular lymph spaces are distended with fibrin and cells and contain very many swine-plague bacteria.

Sections prepared from tissue in which the disease is much more advanced show a complete occlusion of the alveoli and small air tubes with dense cellular masses in which swine-plague bacteria are more or less abundant. The tissue surrounding such foci contains in the alveoli a moderate number of round cells, largely intermixed with fatty cells. Bacteria absent.

No. 10. October 19. Small male pig in poor condition, weighing 21 pounds. Died last night. Subcutaneous inguinal glands very large. On section punctiform hemorrhages and irregular masses of apparently necrosed tissue observed.

In the stomach a considerable number of single and confluent ulcers with projecting, yellowish, friable slough, resting on a firm, indurated, whitish base. The mucosa of small intestine shows extensive arborescent injection of minute vessels.

Contents of large intestine, a chocolate-colored liquid containing much sand. The lesions of the mucous membrane are about as extensive as in case 9. The larger patches show more diptheritic deposit than in preceding case.

Spleen very large, blackish, friable. Cystic degeneration of both kidneys.

Lungs extensively diseased. Nearly the whole of the left lung excepting the caudal half of the principal lobe hepatized, and covered with a thick membranous exudate. This is readily peeled off in shreds and larger patches, especially dense on the ventral and cephalic lobe and adjacent pericardium. Lobes adherent to one another; adhesions easily loosened. The ventral lobe is a mottled bright red. The small air tubes occluded with semi-solid plugs, easily squeezed out. This condition not observed in the other lobes, excepting in the principal. in which the terminal bronchus and branches are occluded with plugs of lung worms and muco-pus.

In the right lung, the ventral lobe in the same condition as corresponding lobe of left side, containing in addition numerous necrotic foci. The cephalic fifth of principal lobe hepatized; the hepatization of a bright red color, with faint regular yellowish mottling. Cephalic half of the small median lobe in the same condition. The bronchi in the tip of principal lobe occluded with masses of lung worms and in part hepatized. In trachea and bronchi much muco-pus.

Bacteriological examination : In preparations from the spleen bacteria not detected. An inclined agar tube inoculated with platinum needle contained in a few days a large number of minute apparently identical colonies. So far as could be determined no hog-cholera bacteria present. A bouillon culture inoculated with a particle of pulp contained both hog-cholera and swine-plague bacteria.

From the liver the same cultures prepared. In the agar tube besides a large number of small colonies are six larger ones. These consist of hog-cholera bacilli, the former of swine-plague bacteria. In the bouillon tube both germs are present.

No inculations or cultures were made with lung tissue from this case. An agar and a bouillon tube were inoculated with a platinum loop from pleural exudate at the autopsy. In both only swine-plague germs appeared.

No. 11. October 22. Black female, weighing 22½ pounds, died last night. Subcutaneous inguinal glands indurated: cortex reddened. Some hemorrhagic points in parenchyma. Fine shreds of exudate on serous surface of large intestine.

In stomach a small quantity of food. Over the entire fundus the mucosa intensely congested and swollen. No ulceration. Small intestine not affected. Large intestine contains a small quantity of turbid fluid. Mucosa much pigmented. In the cæcum and upper colon a small number of ulcers with slightly projecting slough ; lower colon considerably congested. Follicles swollen ; when compressed, a white curdy plug issues from a central opening.

Both lungs diseased. In the right lung near the caudal border of principal lobe a wedge-shaped mass of hepatized tissue, of a bright red color. The peribronchial and interlobular tissue infiltrated over this region, and the pleura covered with a thin exudate. In the bronchi of this lobe are masses of mucus and pus in which lung worms are imbedded. In both cephalic and ventral lobes are masses of hepatized tissue ; in the former also a large necrotic mass.

The two small lobes of the left lung are converted into a greenish-yellow necrotic mass, which cuts like firm cheese. In the principal lobe are three foci of hepatization, two of them near the tip of the lobe, where the bronchus and branches are occluded with lung worms imbedded in mucus.

Pericardium thickened. vessels injected.

Cultures made in bouillon and on agar from pleural exudate. The bouillon culture contained after several days only hog-cholera bacilli. The agar tube was lost. From a particle of lung tissue a gelatine roll A, and agar plate B prepared. On the agar plate six small colonies appeared, made up of swine-plague bacteria. No colonies appeared in the gelatine roll.

A rabbit was inoculated with a particle of lung tissue. It died in 9 days with characteristic hog-cholera lesions. Cultures from the spleen contained only hog-cholera bacilli.

From the spleen and liver, bouillon and agar tubes were inoculated with platinum wire. In all four tubes only hog-cholera bacilli appeared subsequently.

No. 12. October 26. White female pig, weighing about 30 pounds. Died yesterday afternoon and placed in refrigerator until this morning. Subcutaneous inguinal glands enlarged, hyperæmic. In abdominal cavity some dark-colored serum.

One ulcer on mucous surface of lower lip and one on gums. Stomach contracted.

Considerable pigmentation of mucosa. Outside of cardiac expansion two ulcers with slightly projecting slough. Duodenum very deeply pigmented. Pigmentation extends through small intestine. Similar dark pigmentation of the mucosa of large intestine. Peyer's patch at valve converted into numerous small yellowish sloughs with indurated base. A few specimens of trichocephalus attached to mucosa. In upper colon about six small and one larger ulcer with indurated base. Follicles much enlarged with purulent contents easily expressed. Mesenteric glands with very hyperæmic cortex.

Both lungs adherent to surrounding structures. In both pleural sacs a small quantity of clouded serum. On removing the lungs from thorax, the various lobes are found bound together by exudate, the small median lobe imbedded in it and the diaphragm adherent to base of lungs, the intervening exudate being very abundant. In the right lung, both ventral and cephalic lobes, hepatized. On section they have a red, granular appearance mottled with grayish lines and patches. In the principal lobe at its caudal extremity the interlobular tissue distended with reddish serum. Lung worms located here in large numbers. The cephalic portion of this lobe where it rests against the ventral lobe is hepatized, bright red and mottled with whitish points and patches, differing somewhat from the necrotic foci of former descriptions.

In the left lung the ventral lobe solid, resembling the corresponding right lobe. Left principal like the right. Lung worms in the caudal region.

Pericardium thickened, opaque, distended with yellowish serum. The inner surface lined with a partly opaque, partly gelatinous exudate. The heart surface completely covered with a thick, spongy pseudo-membrane. (See plate X.)

Spleen very slightly enlarged and congested. Slight post-mortem changes indicated in the appearance of the liver.

Bacteriological examination: In cover-glass preparations from different parts of the lung tissue very many swine-plague bacteria detected. Two agar plates and two gelatine rolls were prepared. The latter were spoiled. On the second agar plate a large number of minute colonies of the same appearance and three or four miscellaneous colonies. The small ones are swine-plague bacteria.

A rabbit inoculated with a particle of lung tissue died within 20 hours. In the spleen and blood very many swine-plague germs. In a bouillon culture the peculiar growth in minute granules observed which this germ has frequently exhibited heretofore.

At the autopsy an agar and a bouillon culture were inoculated from each pleural sac. In all four tubes only swine-plague bacteria appeared on following day.

From spleen and liver, agar and bouillon tubes inoculated, the former with a platinum needle, the latter with a particle of pulp. On the agar surfaces a small number of isolated colonies appeared, made up of swine-plague bacteria. In the bouillon tubes, besides the clumps of swine-plague bacteria, a large spore-bearing bacillus appeared in bottom of tubes.

No. 13. October 28. Black female pig weighing 28 pounds. Died last night. Subcutaneous inguinal glands enlarged, firm, pale red, mottled with gray.

Mucosa of stomach slightly discolored in fundus, otherwise normal. In duodenum mucosa dark bluish in color, minute vessels injected and occasionally ruptured.

Contents of large intestine a brownish liquid. Slight pigmentation and injection of vessels, no ulceration.

Liver tissue rather firm to the touch. On the surface of the right half yellowish-gray spots, involving from one to several lobules, about one to a square inch. Over the entire surface are disseminated minute dots of a red-lead color, each situated in the center of an acinus. Bile thick and flaky.

Spleen much enlarged, flabby; pulp dark red.

In the left lung the dependent half of ventral lobe solid, grayish red, faintly mottled on the surface. Bronchioles filled with plugs of thick muco-pus. In the central portion of the principal lobe two small hepatized foci. Numerous lung worms in terminal bronchus.

In the right lung between ventral and cephalic lobes a portion of hepatized mottled lung tissue. Somewhat back of this region a mass of necrotic tissue, separated from the surrounding tissue by a greenish-white line and covered by a thick opaque pleura of the same color. In the principal lobe a large wedge-shaped mass of hepatized tissue extending inwards from the lateral border several inches. On section it is dark red, mottled with grayish circles and irregular lines. The pleura much thickened, opaque. Another wedge-shaped hepatized mass situated near caudal tip, perhaps further towards necrosis than the preceding one. Pleura covering it necrosed. Lung worms in the terminal bronchus imbedded in mucus.

Cover-glass preparations from the more recent pneumonic masses show very few germs. From the right pleural cavity a bouillon and an agar tube inoculated. In the former, motile hog-cholera bacilli; in the latter about seven colonies of the same size composed of motile bacilli, probably hog cholera, though they appear somewhat too large. A black rabbit inoculated in subcutis with a particle of lung tissue died in seven days. At the place of inoculation the subcutis is infiltrated over an area of several square inches with pus. No peritonitis; spleen large, containing many hog-cholera bacilli; necroses in liver.

In cover-glass preparations from spleen and liver a considerable number of hog-cholera bacilli detected. From spleen an agar and a bouillon tube were inoculated with platinum needle. Both contain only hog-cholera bacilli on following day. Cultures from the spleen gave the same result.

The lesions found in these thirteen cases may be summarized briefly as follows:

No. 1. October 12. Extensive hepatization of lungs with necrotic foci; lung worms; exudative pleuritis and pericarditis. Congestion and necrosis of stomach; hyperæmia, pigmentation, and ulceration of the large intestine; ulcers small, ulceration evidently follicular. Only swine-plague found.

No. 2. October 12. Spleen large. Fully four-fifths of lungs hepatized, with necrotic foci; pleuritis and pericarditis; lung worms; bronchitis. Small ulcers in large intestine. Only swine-plague bacteria found.

No. 3. October 12. Spleen large. Pneumonia slight; no pleuritis. Hemorrhagic inflammation of stomach. Ulcers in large intestine. Both hog-cholera and swine-plague bacteria detected.

No. 4. October 14. Multiple pneumonic foci in lungs. No pleuritis. Lung worms. Hyperæmia of stomach. Ulcers and patches of necrosis in large intestine. Only hog-cholera bacilli detected.

No. 5. October 15. Scattering pneumonic areas; bronchitis; lung worms. Ulcers in large intestine (button ulcers). Only swine-plague bacteria detected.

No. 6. October 15. Considerable hepatization of all lobes; lung worms. Large ulcers in large intestine. Only hog-cholera bacilli detected.

No. 7. October 15. Spleen large. Hepatization not extensive; pleuritis; bronchitis; no lung worms. Only Peyer's patch near valve ulcerated. Only swine-plague bacteria found.

No. 8. October 17. About four-fifths of lung tissue hepatized, with necrotic foci; pleuritis and pericarditis. Lung worms. No ulcers in large intestine. Infarcts in kidneys. Only swine-plague bacteria found.

No. 9. October 18. About one-half of lung tissue hepatized with necrotic foci; pleuritis; lung worms; bronchitis. Necrotic patches and ulcers in stomach and large intestine. Spleen large. Both hog-cholera and swine-plague bacteria detected.

No. 10. October 19. Spleen large. Three-fourths of lung tissue hepatized, with necrotic foci; pleuritis; bronchitis; lung worms. Ulceration of stomach and large

43

intestine as extensive as in No. 9. Both swine-plague and hog-cholera bacteria present.

No. 11. October 22. Scattering pneumonic foci, with extensive necrosis; pleuritis; pericarditis; lung worms. Congestion of stomach. Small number of ulcers in large intestine, pigmentation. Hog-cholera and swine-plague bacteria present.

No. 12. October 25. Anterior (cephalic) half of lungs hepatized. Extensive pleuritis and pericarditis; lung worms. Pigmentation of mucosa of large intestine. Few ulcers in large intestine; follicles with purulent contents. Only swine-plague bacteria present.

No. 13. October 28. Pneumonia not extensive; pleuritis; bronchitis; lung worms. Intestines pigmented; no ulceration. Only hog-cholera bacilli found.

In all there was pneumonia more or less extensive, associated in the majority of cases with cellular and fibrinous pleuritis, more rarely with pericarditis. In the hepatized regions necrotic masses were frequently met with. Bronchitis was common; lung worms were very abundant. The lesions of the digestive tract varied from case to case. In some the mucosa of the stomach was hyperæmic, bordering on hemorrhage; in others diphtheritic and ulcerated; in still others quite unchanged. The lesions of the large intestine ranged from hyperæmia and pigmentation to the most extensive destruction of the mucous membrane. Of the intermediate stages may be mentioned tumefaction of follicles, with discharge of purulent contents and subsequent formation of small ulcers.

The results of the bacteriological work may be tabulated as follows:

No.	Spleen.	Liver.	Lungs.	Pleura.	Pericardium.	Remarks.
1..	Swine plague.	....	Swine plague.	Swine plague	Swine plague.	No hog cholera.
2..	....do	Swine plague.	....do	....do	Sterile	Do.
3..	Hog cholera ..	....	Swine plague and hog cholera.			Swine plague and hog cholera.
4..	....do	Hog cholera ..	Hog cholera ..	....	....	No swine plague.
5..	Sterile	....	Swine plague.	....	....	No hog cholera
6..	Hog cholera ..	....	(?)	....	....	No swine plague.
7..	Swine plague.	....	....	Swine plague	....	No hog cholera.
8..	....do	....	Swine plague and strepto-cocci.	....do	....	Do.
9..	Hog cholera and swine plague.	Hog cholera and swine plague.	Swine plague.	... do	....	Swine plague and hog cholera.
10..	Swine plague and hog cholera.	Swine plague and hog cholera.	....	... do	....	Do.
11..	Hog cholera ..	Hog cholera ..	Swine plague and hog cholera.	Hog cholera ..	....	Do.
12..	Swine plague.	Swine plague.	Swine plague.	Swine plague.	....	No hog cholera.
13..	Hog cholera..	Hog cholera ..	Hog cholera ..	Hog cholera ..	....	No swine plague.

It will be observed that in six cases (Nos. 1, 2, 5, 7, 8, and 12) only swine-plague bacteria were detected; in three (Nos. 4, 6, 13) only hog-cholera bacteria, and in four (Nos. 3, 9, 10, 11) both were found. In general I would not place too much stress on the absence of either kind of bacteria, because there is no reason why if both pathogenic species are present they should not after a time invade every animal, unless the presence of one disease, such as swine plague, would oppose the invasion of hog cholera in the same animal, a hypothesis for which we have no supporting facts. If we turn to the positive evidence we find that in ten out of thirteen animals the same virulent swine-plague bacteria were found. We have, in other words, bacteria pathogenic in pigs, as we shall presently show, which travel from animal to animal and produce a more or less extensive pneumonia with pleuritis. These facts are in themselves sufficient to separate swine plague as a distinct disease from hog cholera.

In those cases thoroughly studied, such as Nos. 1 and 2, hog-cholera bacilli were probably entirely absent. A glance at the table will show that when they were detected they were always found in the spleen. In Nos. 1 and 2 only swine-plague bacteria were found in the spleen. The same may be said of No. 12. Why the hog-cholera bacilli should have been in these animals and not appear in any of the numerous cultures made is certainly incomprehensible.

Perhaps the best evidence, aside from inoculation, that swine-plague bacteria are the cause of the pneumonia is deducible from the bacteriological examination of the lungs and pleura. In only two later cases were hog-cholera bacilli obtained from the pleura. In one (No. 13), the lung disease had advanced to caseation, and it accords with former experience that in such cases swine-plague bacteria are gone, and if hog-cholera bacilli are in the animal they are certain to appear in these dead lung masses, and hence in the pleural cavity. In those lungs which were extensively hepatized, but in which necrosis had not advanced far, the hepatized tissue was practically a pure culture of the swine-plague bacteria. The cause of the intestinal lesions must remain a matter of doubt, although their nature combined with the presence of hog-cholera bacilli in the herd would lead us to regard them as due to the latter.

As to the origin of this mixed outbreak, nothing positive could be learned of the former history of the herd. As it was made up of two, possibly three lots, it may be assumed that one lot brought one germ and another lot the other. This hypothesis gains force from the great variation in the extent and character of the lung and intestinal lesions. Meanwhile it must be remembered that only one-third of the herd came under observation, owing to the rapidity with which the animals succumbed and the somewhat tardy information concerning the outbreak which reached us.

THE PRODUCTION OF DISEASE IN SWINE AND OTHER ANIMALS WITH THE SWINE-PLAGUE BACTERIA FROM THIS OUTBREAK.

The bacteriological notes already given indicate a virulent variety of these bacteria, inasmuch as rabbits succumbed to a subcutaneous inoculation of a minute dose within 20 hours. The following tests on four pigs were equally striking. The culture used was derived from case 2 by inoculating a rabbit with a particle of lung tissue. The culture from the spleen of this rabbit was again tested about a month later on a second rabbit. The agar culture from this animal was used to inoculate six agar tubes. From these, when 2 days old, the surface growth was washed down into the condensation water with sterile bouillon and the turbid liquid transferred to a sterile test tube. With this the following inoculations were made :

No. 272, 2 cubic centimetres subcutaneously, one-half into each thigh.

No. 273, one-half cubic centimetre into right lung, through chest-wall, between fifth and sixth rib.

No. 274, 1 cubic centimetre into abdomen.

No. 275, 1½ cubic centimetres into right lung.

These pigs were all alike, Essex grade, 10 weeks old, and weighed about 40 pounds each. The inoculations were made November 16, 5 p. m.

No. 272 did not show any ill effects except small tumors at the places of injection. It was killed December 9 and found normal.

No. 273 was very sick on the following day. It breathed with difficulty and refused to rise and eat. On the fourth day there was some improvement. The animal began to eat. The improvement continued and on December 2 (16 days after inoculation) it had apparently fully recovered. On this day it was killed for examination. About 100 cubic centimetres pale yellow serum in abdominal cavity. Diaphragm pressed downwards and very tense, as viewed from the abdominal side. Lungs adherent to chest-wall and diaphragm. Both pleural sacs converted into large abscess cavities, the walls of which are formed by the diaphragm, the chest-wall, and the convex surface of the lungs. These walls are very much thickened and covered with a pulpy yellowish-white layer of pus. The cavity is distended with a turbid, milky fluid. Lungs much compressed, along dorsal border collapsed. Small median lobe collapsed.

Pericardium adherent to heart surface by means of fibrous bands. Pericardial sac contains a small quantity pale yellowish turbid liquid.

Intestinal tract normal. Liver with many lobules either entirely or only centrally congested. Serous surface facing diaphragm slightly roughened.

At this time no swine-plague bacteria were found in the spleen, as the cultures made therefrom with particles of pulp remained sterile.

No. 274 was found dead early next morning, i. e., within 24 hours after inoculation. The superficial inguinal glands much reddened. The abdomen contains about 100 cubic centimetres of clouded, straw-colored serum. The serous surface of the various abdominal organs exposed by reflection of the abdominal walls covered with a thin layer of pale yellowish, friable exudate, thickest on the liver. The exudate also found between the coils of intestines. In both pleural sacs from 50 to 75 cubic centimetres of turbid serum. Pleuritic exudate on the ventral third of the right lung. Similar exudate on the pericardium. Lungs normal. Digestive tract normal so far as the mucous membrane is concerned. The various lymph glands of thorax and abdomen slightly reddened. Liver, spleen, and kidneys not affected.

Only the spleen of this case was examined for bacteria. After scorching the surface an agar and a bouillon culture inoculated with platinum wire. On the agar surface about 100 isolated colonies from 1¼ to 2 millimetres in diameter appeared on the following day, all alike and so far as examined made up of swine-plague bacteria. The bouillon culture likewise proved to be a pure culture of the injected bacteria.

No. 275 became very sick after the inoculation. It refused to get up and eat, and breathed with difficulty. It was found dead November 19, i. e., between 2 and 3 days after the inoculation. Skin not discolored. The cut subcutaneous vessels exude drops of dark blood. In the abdomen a few elastic threads of coagulated exudate stretched across coils of intestines.

In right pleural cavity about 100 cubic centimetres of blood-stained serum. Lungs adherent to ribs by means of a thick, soft, easily removable exudate. Lobes of right lung glued to each other and to pericardium; they are not enlarged. The pleura of the ventral half of principal lobe is converted into a thick, wrinkled, and bleached layer; beneath it the lung tissue is hepatized. The dorsal portion of this lobe, still air-containing, covered with patches and shreds of exudate. The cephalic and ventral lobe solidified, not enlarged; grayish-red on section.

The left lung likewise covered in part with pleuritic exudate. The ventral two-thirds of principal lobe covered with a thick pseudo-membrane and hepatized, very firm; on section, red, mottled with gray. The ventral and cephalic lobes collapsed, covered with exudate, which extends to contiguous pericardium. (See plate VI). The entire diaphragmatic portion of pleura of both principal lobes and median lobe converted into a wrinkled, necrotic mass. Pericardial sac contains much reddish, turbid fluid. The surface of the heart covered with a whitish, firm, roughened exudate.

Stomach contracted; contains a small quantity of bile-stained liquid. The mucosa of the fundus bluish-red, swollen. The whole covered with a layer of viscid bile-stained mucus. The remainder of digestive tract free from inflammation.

In the liver the central region of acini dark brownish red, the outer portion pale brownish.

The presence of the injected swine-plague bacteria in the organs of this animal were determined by the following cultures:

From the spleen, in which no bacteria were detected under the microscope, an agar tube inoculated with platinum wire and a bouillon tube by adding a particle of pulp. The bouillon tube became clouded with swine plague. The agar tube remained sterile. A rabbit inoculated subcutaneously in the ear with a platinum loop dipped into the bouillon culture died within 20 hours. In blood and spleen very large numbers of swine-plague bacteria. In cultures from these organs only these germs present.

A small area on the hepatized left principal lobe scorched, and a small particle cut out with flamed scissors. From the serum filling the cavity thus formed, which contained large numbers of swine-plague bacteria, one gelatine roll, and from this two agar plates prepared. The roll remained free from growth; also one agar plate. On the other appeared a large number of minute colonies and several large colonies of *bacillus subtilis*. The former were identified as swine-plague colonies. From the left pleural exudate an agar tube was inoculated. Abundant growth of swine-plague bacteria on the following day.

From the hepatized region of the left principal lobe portions were placed in alcohol and subsequently sections prepared. The alveoli and small air tubes densely packed with masses of cellular exudate. The interlobular tissue distended with a network of fibrin and a scanty number of leucocytes. In the contiguous, still pervious areas, some air tubes were occluded with cellular plugs, and either around these or independent of them were isolated groups of alveoli occluded with round cells. Swine-plague bacteria were present in enormous numbers both in the parenchyma and the interlobular tissue. They were densely and uniformly sprinkled around and between the cells, in some places forming dense zoöglœa. They were much smaller than in cultures, being just visible at 500 diameters. (See plate XI, figs. 2 and 3.)

These inoculations show that these bacteria injected subcutaneously have little or no effect, but when introduced into one of the large serous cavities severe inflammation is aroused followed by speedy death. In No. 275, the injection into the *right* lung led to a typical pleuro-pneumonia of the *left* lung, such as was encountered in the outbreak from which the bacteria were isolated.

These bacteria were fatal to mice and guinea-pigs and to pigeons in large doses :

December 13.—Two mice inoculated under skin of back with a loop of growth from an agar culture one day old. Both dead next morning, with spleen much swollen and containing large numbers of the inoculated bacteria. From the spleen of one an agar and a bouillon culture were made. The bouillon became faintly clouded ; on the agar numerous confluent colonies appeared. Only swine-plague bacteria detected in these cultures.

From the same culture a guinea-pig was inoculated subcutaneously in the same way. The guinea-pig lived 12 days. No lesions except a fatty condition of liver found at autopsy. A tube of agar to which 2 drops of blood were added remained sterile.

At the same time two fowls and two pigeons were inoculated, one fowl and one pigeon with a loop of the agar culture subcutaneously and one fowl and one pigeon with 1 cubic centimetre of a peptone bouillon culture one day old. The needle of the syringe penetrated the superficial muscular layer of one pectoral muscle.

Both fowls remained well, also the pigeon inoculated with the loop. The other pigeon dead next day. On the right pectoral region the subcutis was infiltrated and of a reddish-yellow color, the subjacent muscular tissue discolored to a depth of one-fourth to one-half inch. Lungs congested, other organs normal. In the blood numerous swine-plague bacteria giving the polar stain ; in the liver very few.

In April, 1891, about 1½ years after these bacteria had been obtained from this outbreak, the following inoculations were made to determine how much of their virulence had been lost by cultivation.

April 13.—With a loop rubbed over an agar culture 4 days old, a large gray rabbit was inoculated by inserting the loop into an incision on the ear made with flamed lancet. Rabbit dead next morning. In blood and spleen swine-plague bacteria in moderate abundance.

April 21.—Two guinea-pigs weighing 1½ pounds each received a subcutaneous injection of 0.075 and 0.15 cubic centimetres of a turbid suspension of swine-plague bacteria from an agar culture 1 day old.

The guinea-pig which had received the largest dose died in 26 hours. At the place of injection slight yellowish infiltration of subcutis. No peritonitis. In both pleural sacs a large quantity of a grayish, partly gelatinous effusion, containing immense numbers of bacteria. Pericardial sac similarly involved. Both lungs dark bluish red throughout; tissue still floats.

The other guinea-pig very sick for several days, recovered, and was killed on the eighth day. At the place of injection a minute ulcerous opening from which pus could be pressed. In the subcutis considerable rather firm infiltration, surrounded by patches of blood extravasation. Internal organs not affected.

VIII.

In November, 1889, an outbreak of swine plague came to the notice of the Bureau, which, in some respects, has a quite remarkable history. The information given below was obtained by Dr. Kilborne.

An educational institution near this city was in the habit of keeping on its inclosure a small number of swine in pens which were arranged in the manner indicated in the figure. The slope of the ground was from yard 1 to yard 3, and from yard 5 to yard 3.

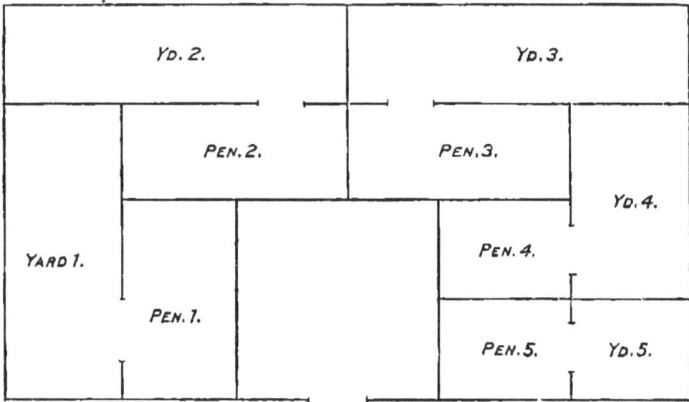

In September a boar, 5 months old, died in pen 1. There had been no disease on the place 2 years previous to this occurrence, and no recent purchase of pigs. In the latter part of October ten pigs, 2 months old, were purchased, five placed in pen 1, and five in pen 5. At this time a sow in pen 2 had four pigs, all of which died in a few days, while the sow, although sick for a time, recovered.

In pen 3 a sow became sick very suddenly November 8, and died next day. A litter of seven pigs, 7 weeks old and small f " their age, died within 4 days after the death of the sow. Of these, five came under our observation (Nos. 1 to 5 inclusive, of the autopsy notes). These five, after a very careful examination, proved to be cases of swine plague. While the recently purchased pigs in pen 1 remained unaffected, those in pen 5 began to die at the same time that the young pigs in pen 3 were dying. Thus two died November 9 and one November 10. One of these was carefully examined, and to our surprise the disease was found to be hog cholera, as the autopsy notes (No. 6) clearly show. The remainder were taken to the Experiment Station. At the same time there were in pen 4 two pigs about 5 months old and one old sow. One of the pigs, sick November 12, was transferred to the Experiment Station, where it was found dead on the following morning. This case also was one of hog cholera (Case 7). By feeding the viscera of these two pigs to fresh pigs an outbreak of hog cholera was produced, which was subsequently utilized in a series of experiments on vaccination, as a means of exposing swine which had been vaccinated beforehand.

On inquiry over a month later, we were informed that no further losses from swine diseases had been sustained.

In the following pages are given the notes of the autopsies and bacteriological examination of the litter of small pigs which died in pen 3. They are all the more interesting in that the disease was swine plague uncomplicated with hog cholera. The autopsy notes of the two hog-cholera cases from pen 5 and pen 4 are appended, to complete the history of this remarkable outbreak.

November 11, 1889.—Pig No. 1, white female, weighs about 15 pounds. Died last night. Considerable reddening of the skin over the ventral aspect of the body.

Stomach contains a small quantity of a yellowish viscid liquid. Fundus covered with mucus and intensely reddened and swollen. In the large intestine, mucosa slightly discolored ; in a few places somewhat reddened. No necrosis or ulceration. Glands of the meso-colon enlarged, indurated, pale.

About one-half of both lungs hepatized, the disease limited to the ventral or dependent half. The various lobes adherent to one another, to pericardium and diaphragm by means of a thin, papery pseudo-membrane, which is removable. The hepatization is very firm ; on section, grayish red or red mottled with gray. Near the roots of the lobes, also on diaphragmatic surface of the lungs, are considerable numbers of small necrotic masses, surrounded by a bluish zone. In the trachea and bronchi, reddish froth ; in the distal extremities of both bronchi, lungworms imbedded in mucus.

Spleen small, not congested. Liver exceedingly firm to the touch. Surface not smooth, owing to acini slightly projecting above interlobular tissue.

Bacteriological examination : Preparations from hepatized lung tissue contain an immense number of very minute oval bacteria. Those from the pleural exudate contain a smaller number. On one of the principal lobes the pleura was scorched over the diseased portion, and with a particle of tissue from within the scorched area, a gelatine roll A, and two agar plates B, prepared. In the gelatine roll appeared, after a few days, a large number of barely visible colonies. A week later, besides these, a smaller number of colonies, 4 to 5 times larger than the preceding, had appeared. Owing to the large number of colonies microscopic examination not successful. Minute bits of gelatine were removed at different intervals with warmed platinum wire and transferred to peptone bouillon. In these tubes only swine-plague germs appeared. The larger colonies may have been streptococci, as they failed to develop in the bouillon. Both agar plates, after several days, contained a large number of identical quite small colonies, which were found to be swine plague colonies by microscopic examination and transfer to bouillon. In order to determine whether any other bacteria were present, two bouillon tubes were inoculated *directly* from the lungs, by piercing them with a platinum wire and transferring this to the bouillon. In both tubes *only swine-plague bacteria* appeared.

A rabbit inoculated at the same time with a particle of lung tissue died in 5 days. The subcutaneous tissue over abdomen extensively thickened by a purulent and gelatinous infiltrate extending over part of thorax. The abdominal walls are thickened and glued to the viscera. These latter covered with a rather firm elastic exudate, which dips down between the coils. The serosa is sprinkled with punctiform ecchymoses. Spleen small ; thoracic organs not affected. In the abdominal exudate are immense numbers of oval bacteria, staining rather feebly. These not detected in spleen and blood. An agar tube inoculated from the spleen contained but a single swine-plague colony. A gelatine culture from the blood shows in the track of the needle about 12 minute round colonies. In a bouillon culture from the peritoneal exudate only swine-plague bacteria developed.

From the pleural exudate of the pig an agar and a bouillon tube inoculated at the autopsy. In the former a considerable number of isolated and confluent colonies of swine-plague bacteria appeared. The bouillon culture likewise contains only swine-plague bacteria.

From the spleen pulp, in which no bacteria were detected, a particle placed in an agar and a bouillon tube. Both were sterile on the following day. On the third day faint cloudiness of the bouillon, which contained only swine-plague germs. In the agar tube the condensation water was clouded, and a grayish membrane starting from it ascending along agar surface. Only swine-plague bacteria detected in this growth.

Portions of lung tissue from different lobes, hardened in alcohol, were examined. Sections from recently affected lung tissue, bordering on normal tissue, showed the alveoli to be filled in some lobules with blood corpuscles and fibrin; in others there was, in addition, a filling up of scattered alveoli with round cells. In still others the alveolar capillaries were greatly distended with corpuscles, almost occluding the alveoli. Imbedded in the fibrinous plugs of the alveoli were colonies of minute cocci, almost every alveolus containing one or more such colonies. In sections from tissues in more advanced stages there were, in addition to the more dense cell infiltration, large masses of the minute bacteria occupying the alveoli in some portions of the section. Any regularity in the distribution of these bacteria not observed.

No. 2. Examined at the same time. White female, weight 15 pounds. Skin on ventral aspect of body moderately reddened. Subcutaneous inguinal glands hyperæmic.

Stomach contains a small quantity of turbid liquid. Mucosa of fundus considerably reddened. In the small intestine the vessels of villi appear injected, especially in duodenum. In the cæcum and colon the entire mucosa has an intense purplish hue, shading into a wine color. This most marked in the cæcum and upper 4 or 5 inches of colon, where the epithelium appears necrosed. The inflammation gradually diminishes and disappears in the rectum.

Exudative pleuritis as in case 1; the pseudo membrane as thick as heavy paper. About two-thirds of right lung hepatized. The cephalic and ventral lobe entirely solidified, also adjoining half of the principal lobe. In the caudal portion of the latter several hepatized foci. Lung worms in terminal bronchus. The tip of cephalic lobe completely necrosed. The ventral lobe contains large, yellowish-white, homogeneous foci of dead tissue. Median lobe completely hepatized. Through it are disseminated necrotic foci.

In left lung, principal lobe entirely hepatized. Hemorrhagic, grayish-red, and grayish lobules found on the same cut surface. Ventral and cephalic lobes merely congested. Pericardium thickened.

Spleen small, pulp darker than normal and softened. Liver as in No. 1. Bile very thick, dark-colored.

Bacteriological examination : An agar and a bouillon tube were inoculated each with a particle of spleen pulp. The agar tube remained sterile. The bouillon became faintly clouded on the second day and contained only swine-plague bacteria.

The hepatized lung tissue and pleural exudate both show presence of swine plague germs; the lung tissue contains immense numbers of them. With a bit of the latter a gelatine roll A and agar plate B prepared. In the gelatine roll, a considerable number of very minute brownish colonies appeared within a week. They were all alike. Several transferred to bouillon at intervals and the resulting cultures carefully examined. Only swine-plague bacteria detected. The agar plate had a moderate number of colonies, those growing on surface from 2 to 3 millimetres in diameter. These also proved to be swine plague when examined and transferred to bouillon for further identification.

At the autopsy an agar and a bouillon tube were inoculated from the right pleural cavity. On the agar appeared a large number of isolated and confluent colonies all apparently alike. Some of these, as well as the growth in the condensation water, were examined and found to be swine-plague bacteria. The bouillon contained also a streptococcus and a bacillus, imparting a sour smell to the culture. A rabbit inoculated with a particle of lung tissue died in 6 days. The subcutaneous infiltration and the peritonitis precisely as in the rabbit inoculated from No. 1. The spleen contained scarcely any, the blood few, and the peritoneal exudate an immense number of swine-plague bacteria. Agar cultures from exudate and blood and a bouillon culture from the exudate contained only swine-plague bacteria.

The presence of swine-plague bacteria on the inflamed mucosa of cæcum was demonstrated by inoculating a rabbit with a particle of mucosa which had been washed in sterile water. Rabbit died in 2 days with considerable local infiltration, but no peritonitis. An agar culture from spleen with platinum needle remained sterile. A bouillon culture from blood became clouded with swine-plague bacteria. To further test this germ, one-eighth cubic centimetre of this bouillon culture 7 days old was injected subcutaneously into a rabbit. It lived 11 days. There was extensive purulent infiltration of the subcutis over abdomen and thorax. Internal organs normal. But one (bouillon) tube inoculated with particle of liver tissue. This remained clear. No. 3 examined on same day. White female weighing 18 pounds. Died last night. Ventral aspect of body considerably reddened. Stomach as in preceding cases. Mucosa of duodenum very much reddened; the remainder of the small intestine normal. Mucosa of cæcum and colon not quite so intensely inflamed as in No. 2. Indications of a yellowish, soft exudate appearing in small masses as part of the feces. Its microscopic characters not determined.

Double exudative pleuritis, the exudate, thick on diaphragm which firmly adheres to lungs, very slight on convex surface of lungs where it appears as a roughening or a very delicate membrane. The various lobes glued to each other and to pericardium.

In the left lung the two smaller lobes (ventral and cephalic) in a condition of pale-red hepatization; very slightly enlarged. In the principal lobe four or five foci of dark-red hepatization imbedded in normal tissue, on section marked with grayish, irregular lines. These masses are elevated slightly above the surrounding tissue and covered with a thick, opaque, greenish-white pleura. In the right lung both smaller lobes hepatized, larger than the corresponding left lobes, very firm to the touch. The cut surface sprinkled with minute grayish masses. In the principal lobe the hepatized masses are dark red in color. The small median lobe enveloped in exudate and hepatized, the cut surface grayish red.

In the bronchi a small quantity of reddish, frothy liquid. In the right terminal bronchus lung worms. Bronchial glands very large, firm, yellowish white.

Pericardium inflamed, opaque. In left heart a firm washed clot imbedded in a dark, soft coagulum. The right heart distended with a dark, soft coagulum.

Spleen small, dark-colored. Liver and bile as in No. 2.

Bacteriological examination: From the spleen an agar and a bouillon tube inoculated with a particle of pulp. After several days a grayish growth, spreading from bit of spleen, composed of rather large cocci. The bouillon tube at this time also clouded; contains only swine-plague germs.

Two similar cultures prepared from the liver. In the bouillon tube only swine-plague bacteria appeared. On the agar surface three colonies present, one of these a chromogene, the others large spore-bearing bacilli. In the turbid condensation water the same large bacilli and swine-plague bacteria intermingled. In cover-glass preparations from spleen and liver no germs could be detected.

From the right pleural cavity an agar and a bouillon tube inoculated. Both remained sterile.

From the most recently hepatized lung tissue, which contains large numbers of swine-plague bacteria, a gelatine roll A and an agar plate B prepared. In the gelatine roll colonies appeared answering to the description given for those under case 1. The bouillon tubes prepared from the colonies contained no hog-cholera germs at any time. Swine-plague bacteria and possibly streptococci were present, although this is somewhat doubtful. The agar plate B remained free from growth. A bouillon tube inoculated with a platinum needle thrust into the hepatized lung remained sterile.

A rabbit inoculated with a particle of lung tissue succumbed to the disease in 8 days, with extensive subcutaneous purulent infiltration over abdomen. No peritonitis. Cover-glass preparations and cultures from blood and spleen are negative.

From this lung sections were cut from portions of one principal lobe hardened in alcohol. The alveoli were nearly all occluded by round cells, among which in a certain number of alveoli large masses of the very minute swine-plague bacteria could be detected. The small air tubes likewise filled with densely packed cell masses. The interlobular tissue in a state of inflammatory œdema.

November 12.—No. 4, small female, weighing about twenty pounds. Died last night. Ventral surface of the body moderately reddened. Considerable quantity of subcutaneous fat. Lymphatic glands in groin slightly enlarged and reddened.

One ulcer at base of left lower front tooth. Stomach contracted and contains a small quantity of liquid resembling tomato juice. The mucosa of fundus over an area 3 inches in diameter of a dark wine-red color; the hyperæmia extends through entire mucous layer. On the surface a very delicate, easily removable pseudo-membrane. Mucosa of duodenum pigmented and bile-stained. In lower ileum some patches of punctiform ecchymosis. In large intestine feces adhering rather firmly to mucosa, which is reddened and pigmented in spots and patches and somewhat rough to the touch.

In thracic cavity lungs covered with false membrane and in part adherent to chest-wall and pericardium. On removing them, the pleuritis and hepatization found nearly as extensive as in the preceding case, i. e., the greater part of both small lobes in each lung hepatized and exceedingly firm; in the principal lobes disseminated foci of hepatization both of the grayish-red and the hemorrhagic type. Lung worms not detected. In the air tubes of the ventral lobes cylinders of whitish pus. In the large bronchi reddish frothy liquid. Pericarditis as in preceding case. On the epicardium a very delicate pseudo-membrane.

Liver tissue very firm to the touch; bile thick. Pyramidal portion of kidneys dark red. Spleen small, somewhat darker colored than in normal condition.

Bacteriological examination : Cover-glass preparations of the pleural exudate show a moderate number of swine-plague bacteria. In the lung tissue there are immense numbers of these bacteria, with an occasional large bacillus amongst them. From the right pleural cavity an agar and a bouillon tube, from the left an agar tube inoculated at the autopsy. In these tubes a considerable number of identical colonies appeared made up of swine-plague bacteria. In the condensation water of an agar culture from right pleura occasional streptococci to be seen. The bouillon tube became uniformly clouded with swine-plague bacteria. After two weeks a few very large colonies of strange bacteria had developed in both agar tubes.

With a particle of lung tissue a gelatine roll A and an agar plate B prepared. Roll A broke ; plate B showed in a few days about 50 colonies, evidently alike. Those examined and transferred to bouillon consisted only of swine-plague bacteria.

A rabbit inoculated with a particle of lung tissue remained unaffected. Another rabbit received a subcutaneous injection of one-eighth cubic centimetre bouillon from an agar colony of lung plate. This rabbit died in 13 days with a very large abscess at the point of inoculation.

In the spleen of pig no bacteria were detected with the microscope. In the liver several germs resembling swine-plague bacteria were observed. With particles of liver and spleen tissue two agar and two bouillon tubes were inoculated. Only the bouillon tube from the liver became fertile and contained a diplococcus; no swine-plague bacteria.

Sections from the lung tissue hardened in alcohol and stained in various ways were carefully examined. In the same section were lobules in which the alveoli contained fibrin and very few cells, others in which much desquamation of the epithelial cells had taken place, and others in which the alveoli were occluded by dense cell masses. In some places the septa had apparently disappeared and a continuous plug of densely massed cells extended through a number of continuous alveoli. The small air-tubes were likewise filled up with cell masses. The interlobular tissue distended with serum, the lymph channels similarly distended and containing masses of leucocytes. In all alveoli excepting those containing only the desquamated cells, the very minute oval swine-plague bacteria are present in immense numbers, disseminated singly through the œdematous tissue and in large zoöglœa amongst the cellular masses.

November 13.—No. 5, small white male, weight about 10 pounds. Died last night.
In the digestive tract of this case nothing abnormal excepting a somewhat bluish coloration of the mucosa of large intestine and swelling of the solitary follicles, the contents of which can be expressed through a central opening.
About one-third of the entire lung tissue hepatized. The disease involves the ventral third of all the lobes, excepting the left cephalic, which is free from pneumonia. The median lobe completely solidified and containing two necrotic foci. A thick pseudo-membrane covers the pleura of the diseased areas. In the bronchi and branches of the principal lobes numerous plugs of lung worms imbedded in mucus. The bronchi of the ventral lobes occluded by cylindrical masses of mucus and pus.
Bacteriological examination : An agar tube inoculated from the right pleural cavity at the autopsy contained subsequently 5 colonies of swine-plague bacteria.
Cover-glass preparations of diseased lung tissue show large numbers of pus corpuscles, amongst these small oval bacteria, probably swine plague, and occasional chains of streptococci. A gelatine roll culture A and an agar plate B were made from particles of lung tissue. In roll A there appeared, after a week, about one hundred colonies of *bacillus coli*, twelve colonies of a slowly liquefying chromogene, and a very large number of colonies just showing a disk under the microscope. Particles of gelatine containing these minute colonies were removed to bouillon, but no development took place. On the agar plate about ten colonies of swine-plague bacteria appeared, which grew quite large, owing to isolated position. Transferred to a bouillon their swine plague nature confirmed.
With a particle of lung tissue a large black rabbit inoculated. Dead in five days. The subcutis was extensively infiltrated and thickened as in previous cases. Peritonitis absent. Spleen barely enlarged. In cover-glass preparations from spleen and liver no bacteria observed. In an agar culture from blood thirty swine-plague colonies appeared. After a week a fleshy, whitish growth composed of large motile bacilli starts from condensation water upwards on agar surface. In an agar culture from the liver-a considerable number of swine-plague colonies appeared. In a bouillon culture from spleen only swine-plague bacteria developed.
With particles of spleen pulp of the pig an agar and a bouillon tube inoculated. The bouillon became very turbid, greenish, fluorescent and contained a small motile bacillus. Plate cultures from this proved the bacillus a liquefying fluorescent bacillus. In the agar tube a growth started from particle of spleen down the inclined surface and subsequently imparted to the agar a greenish tint. The same bacillus as that found in the bouillon culture present. No swine plague bacteria detected.
In all five pigs specimens of trichocephalus were attached to mucosa of cæcum and upper colon.

The following tables give a brief summary of the facts obtained:

Pig No. 1. November 11, one-half of lungs hepatized ; some necrotic foci. Exudative pleuritis, pericarditis. Lung worms. Catarrhal inflammation of stomach. Only swine plague bacteria present.

Pig No. 2. November 11, two-thirds of lungs hepatized; many necrotic foci. Pleuritis, pericarditis. Lung worms. Hyperæmia of stomach. Intense hyperæmia of large intestine.

Pig No. 3. November 11, one-half of lungs hepatized ; pleuritis; pericarditis; lung worms; hyperæmia of stomach and large intestine.

Pig No. 4. November 12, one-half of lungs hepatized ; pleuritis and pericarditis ; bronchitis ; hyperæmia of stomach, of large intestine slight.

Pig No. 5. November 13, one-third of lungs hepatized ; pleuritis ;

bronchitis; lung worms; swelling of solitary follicles of large intestine, contents expressible.

Pig No.	Lungs.	Pleura.	Spleen.	Liver.	Intestines.
1	Swine plague .	Swine plague	Swine plague .		
2	... do .... -	...do	...do		Swine plague.
3	...do	Negative	...do	Swine plague .	
4	...do	Swine plague and streptococci.	Negative	Negative	
5	Swine plague, bac. coli, chromogene.	Swine plague	...do	...do	

It will be observed that these five animals died of swine plague pure
and simple. Hog-cholera bacilli were absent. An examination of the
pathological notes shows in every case extensive pneumonia, accompanied by exudative pleuritis and terminating in some cases in necrosis
of lung tissue. Perhaps the most instructive feature of the disease
is the inflammation of the stomach and large intestine, which was
especially marked in the second case. In the third, besides the intense
hyperæmia, there was evidence of exudation.

That such intense hyperæmia, provided the bacteria continue to act
on the membrane, or provided they are of the proper degree of virulence, may lead to croupous and diphtheritic deposits and subsequent
ulceration, needs no comment. Moreover, the swelling of the solitary
follicles with discharge of contents, as observed in No. 5, may lead to
subsequent ulceration. It is interesting to note that the disease reached
its severest expression in No. 2, both as regards lungs and intestines,
and in later cases the lesions became less extensive.

The table giving the results of the bacteriological work shows a gradual disappearance of the swine-plague bacteria and the invasion of other
bacteria. Thus, in the spleens of Nos. 1, 2, and 3, swine-plague bacteria
were detected while the cultures from the spleens of Nos. 4 and 5 remained sterile. That these bacteria perish very speedily in the body
is highly probable when we bear in mind that they die very rapidly in
culture media, a fact to be pointed out further on.

The swine-plague bacteria causing this outbreak were not so virulent
as those of the preceding one as demonstrated by the inoculation of
rabbits from lung tissue and pure cultures. While those of the preceding outbreak destroyed rabbits in 20 hours, these were fatal in from
2 to 13 days, according to the size and age of the rabbit. The following
inoculation, made about 4 weeks after the last case was examined, serves
as an additional illustration :

December 6, 1889.—A large white rabbit inoculated subcutaneously on right ear
with a loop rubbed over the growth of an agar culture. The animal was found dead
December 10. From the place of inoculation on the right ear, suppuration had extended down over the muscles of the neck. In right pleural sac, on pleura of ribs
and right lung, a thick creamy deposit. A similar, more consistent exudate on epicardium, which is more or less ecchymosed. Other organs not affected.

After constant cultivation of this germ for a year and a half on agar, it had lost its virulence to a large extent, so that subcutaneous inoculation produced merely a local abscess. While small doses injected into the circulation failed to kill a rabbit larger doses were still rapidly fatal.

March 31, 1891.—A rabbit receives under the skin of abdomen 0.2 cubic centimetre of a bouillon culture prepared from an agar culture one day old. The rabbit subsequently seemed slightly ill, but recovered. It was chloroformed on the thirteenth day. Over an area several inches square the subcutis is thickened by purulent infiltration, and the skin gangrenous and very dry and hard. The abdominal muscles slightly ecchymosed. Internal organs not affected.

April 13.—Large white rabbit receives into ear vein 0.3 cubic centimetre of a bouillon culture 24 hours old. The rabbit showed signs of slight illness for a time. It began to grow thin, and after several weeks it was noticed to move with great difficulty. Chloroformed on the nineteenth day. Very anæmic and thin. Internal organs pale but normal, excepting kidneys, of which cortex is fatty. Several abscesses under the fascia of the left knee joint, containing a bluish milky pus. Gland in right axilla converted into an abscess.

April 29.—From an agar culture 4 days old a turbid suspension of the condensation water was injected into an ear vein of two rabbits; one received one-fourth cubic centimetre, the other one-half cubic centimetre.

Both very quiet and drowsy on the following day; breathing accelerated; one with head drawn backward.

May 1.—Both dead this morning. In the first animal the head partially drawn backwards on neck. Spleen dark, large, and softened. No peritonitis. Liver very pale; on left lobes much interlobular pale yellowish mottling. Cloudy swelling of cortex of kidneys. Fatty degeneration of heart muscle. Right heart distended with dark, partly coagulated blood; in left, a small quantity. Both lungs œdematous; right hypostatic. In spleen and blood a moderate number of bacteria showing polar stain very clearly. Cultures from spleen and blood contain only the injected bacteria.

In the second rabbit same position of head. Peritonitis indicated by a viscid exudate on cæcum and stretching between coils of intestine when these are separated. On liver and spleen a pseudo-membrane. The exudate consists of endothelium in state of fatty degeneration, strands of fibrin, leucocytes and large numbers of bacteria showing polar stain. Spleen, liver, kidneys, and blood contain no bacteria visible in cover-glass preparations. Liver and kidneys hyperæmic. Heart in diastole. Both sides contain dark, partly coagulated blood; heart muscle fatty. Lungs, especially ventral portions, congested. Cultures the same as in preceding case.

While there existed the difference in pathogenic power between the bacteria of this and the preceding outbreak, none could be detected from a biological and morphological standpoint.

The question naturally arises, why should such attenuated bacteria as these produce such a severe disease? The age of the infected animals probably will answer this question. Observers have not infrequently noted the fact that bacteria sufficiently attenuated to be harmless to old animals prove fatal to the young. The Pasteur school has made use of this observation in increasing the virulence of pathogenic bacteria by first passing them through young animals. It is claimed that after a number of inoculations these bacteria become fatal to older animals as well.

Having thus demonstrated an outbreak of pure swine plague among the animals in pen 3, let us turn briefly to the diseased swine in the other pens which came under observation. It will be remembered that one pig from pen 5 died November 10, and was taken to the Experiment Station, where it was kept in the refrigerator until November 11. The subject is sufficiently important to warrant the reproduction of the notes:

No. 6. Stomach contracted : contains a small quantity of food. One ulcer at the margin of the œsophageal expansion, the latter covered with a thin brownish-yellow layer of desquamated epithelium. In the ileum small ulcers, about 2 to a square inch, one-eighth inch across; adherent slough stained yellow. Extensive ulceration in cæcum and upper colon. The ulcers vary from one-eighth to one-half inch in diameter, blackish on the surface. The base consists of a firm yellowish-white tissue extending into muscular coat. Six inches below the valve one large, button-shaped ulcer, the firm base extending to serosa and three-eighth inch thick. The serosa under it discolored and the meso-colon adherent to it. In the lower colon are a large number of circular pale spots on the discolored, pigmented mucosa, representing probably the places to which exudates had been attached but now shed.

Lymph glands in lesser omentum and meso-cæcum with cortex hemorrhagic.

In left pleural cavity large patches of blood extravasation under pleura of ribs. No pleuritis. Posterior mediastinal and bronchial glands hemorrhagic. Throughout the entire lung tissue are disseminated hepatized foci of various sizes surrounded by healthy and more or less emphysematous tissue. Thus in the right lung about one-half of both cephalic and ventral lobes and a small volume of principal lobe involved. In the left lung one-half of cephalic lobe and tip of ventral lobe hepatized; in the principal lobe six small foci. One-half of median lobe hepatized. The small air tubes plugged with thick muco-pus. In the principal lobe several of the foci directly traceable to plugs of lung worms, which are very abundant.

Spleen somewhat enlarged, discolored ; pulp soft.

Bacteriological examination : In spleen pulp a considerable number of hog-cholera bacteria detected. An agar and a bouillon tube inoculated with platinum wire. In the former a considerable number of isolated colonies of the same size appear, which are, so far as examined, hog-cholera bacilli. In the bouillon tube only hog-cholera bacilli observed. A rabbit inoculated with a particle of spleen pulp died in 4 days. At the point of inoculation a small abscess. Spleen large, dark, softened ; contains a a few hog-cholera bacilli. Numerous very small points of necrosis on liver surfaces. In an agar culture from the spleen only hog-cholera bacilli detected.

With a particle of lung tissue in which a few hog-cholera bacilli were observed a gelatine roll A and an agar plate B prepared. In the roll there were present at end of a week a large number of small brownish colonies and about fifty several times larger. By carefully removing the smaller ones to bouillon these were found to be hog-cholera bacilli. The large colonies not examined. On the agar plate a considerable number of vigorous hog-cholera colonies appeared in 2 days.

Case 7. November 13. Large white female, 5 months old, weight 65 pounds. Ventral aspect of body and ears bluish red. Subcutaneous fat abundant. Lymphatic glands of groin with surface purplish and parenchyma mottled red and gray. A small number of ecchymoses on abdominal side of diaphragm. Spleen very large, blackish, soft.

Stomach partly filled with food. Mucosa of fundus, and, in fact, of fully one-third the whole area of stomach surface, intensely reddened. The hyperæmia extends to submucous tissue. Slight ecchymosis and pigmentation of duodenum. Occasional petechiæ in ileum; Peyer's patch near valve pigmented. Entire mucosa of cæcum and colon purplish gray and dotted with petechiæ. On mucosa of cæcum about 30 pale circular spots slightly depressed, which evidently were the seat of adherent sloughs or exudates. In rectum petechiæ ; glands of meso-colon hemorrhagic.

In the principal lobe of left lung a bright red hepatized mass about three-quarters of an inch in diameter; in the right corresponding lobe a mass half as large. In the bronchi of these lobes very many lung worms and a considerable quantity of reddish gelatinous mucus. No other hepatization and no pleuritis observed. On the surface of the lungs a few petechiæ. The location of the hepatizations makes it probable that lung worms were the original cause.

Liver shows marked post-mortem changes. Kidneys with cortical portion dotted with numerous hemorrhagic points. Clots of blood surrounding papillæ.

In the spleen were observed hog cholera and large bacilli. In an agar culture with wire numernos hog cholera colonies appeared.

It is a remarkable fact that there should be two diseases of different nature in adjoining pens at about the same time. It is not improbable that hog cholera was brought with the ten pigs, but in this case it is quite curious that those in pen 1 were not attacked. We must assume that perhaps but one pig was chronically infected (old ulcers) and that this pig placed in pen 5 formed the starting point of a slowly developing infection among the others.

IX.

In July, 1890, the attention of the Bureau was called to an outbreak of swine disease of a very fatal character about 2 miles northwest of Pleasantville, New Jersey. Veterinarian F. L. Kilborne visited this locality July 21 and obtained the following information from Mr. Joseph Young, the owner of the diseased herd, who gave us all the facilities in his command for the investigation of this plague.

On May 9, 1890, Mr. Young bought forty-five pigs, 2½ to 3 months old, collected by a dealer from several counties. May 23, a second herd of fifty-two pigs was obtained from the same source. In this lot were noticed several unthrifty animals, and coughing was heard up to the time of the appearance of the disease. About July 1 they began to die, one of the unthrifty ones being the first to succumb. Deaths, at first few in number, increased until the number reached from three to eight per day. Thus from July 1 to July 19 thirty-four perished. On July 20 four died, on July 21 seven died, on July 22 eight died.

No swine had been on this farm for several years except four animals, which had been raised and fattened last season and which had always been in good health.

The herd was watered from a well, the waste being permitted to run into a little depression or wallow in a dry sandy soil. The food consisted of slops, brought over from Atlantic City in boats. Several other herds in the neighborhood were being fed in the same way without any untoward results at the time. It should be stated, however, that later several other herds in the neighborhood became infected. This information did not reach us until after the disease had died out.

The symptoms noted by the observant owner were coughing, loss of appetite, and occasional vomiting. The animals strayed into bushes and other hiding places, soon became lean and gaunt and weak in the hind quarters. They died 3 to 5 days after showing signs of disease. Some lived longer, others died quite suddenly.

Of the forty-two pigs remaining July 23, twenty died between July 23 and July 28, five died on the Experiment Station, and only seven, out of a total of ninety-seven, lived through the disease.

Seventeen animals came under our observation. Of these, twelve were examined on the farm and the remaining five sent alive by express to the Experiment Station. Here they all died within a week after their arrival. Of those examined on the farm, Nos. 1 to 4, inclusive, and Nos. 11 and 12, came under Dr. Kilborne's observation. From Nos. 5 to 10, inclusive, cultures were made on the farm by the writer. Cultures were therefore made from eleven cases only.

The bacteriological examination of Nos. 5 to 10, inclusive, can not be regarded as thorough, owing to the innumerable insect pests on the farm. Much work, however, was bestowed upon Nos. 13 to 17, inclusive, which died at the Station. The notes of this outbreak are reported in detail, while the usual summary will be given farther on.

No. 1. July 21, 1890. Weight of animal about 50 pounds. Skin not reddened. A few delicate fibrils of exudate stretching over coils of intestines. Spleen very much enlarged, congested.

Digestive tract. Stomach contains a small quantity of food. Mucosa of fundus intensely hyperæmic, bordering on hemorrhage.

Small intestine nearly normal. Contents of large intestine liquid. The mucosa of cæcum and entire colon nearly concealed by a layer of dirty, greenish-yellow, diphtheritic membrane, quite readily scraped away, exposing a deeply reddened, raw surface.

Right lung normal. The entire left lung is covered with a pale yellowish (friable) exudate, which glues this lung to surrounding parts. The lung itself contains masses of hepatized tissue, aggregating, perhaps, one-fourth of the entire lung.

The lymphatic glands generally are enlarged; the cortex, and sometimes medullary portion also, infiltrated with blood. Other viscera not markedly changed.

No. 2. July 21, 1890. Weight of animal about 75 pounds. Died very suddenly during the day. No skin lesions. Spleen barely larger than normal.

Digestive tract. Only the large intestine was markedly affected. The mucosa extensively pigmented with occasional patches of hyperæmia. In colon just below the valve one ulcer one-half inch across and a few smaller ones discovered.

Each pleural sac contained nearly 500 cubic centimetres of slightly-clouded serum containing large masses of gelatinous coagula. The lungs free from pneumonia. Lymphatic glands and remaining viscera not diseased.

No. 3. July 22. Weighs about 40 pounds; died during the night. Had been sick at least a week according to statement of owner. Spleen moderately gorged with blood.

Digestive tract. The mucosa of cæcum and upper half of colon more or less pigmented and beset with rather deep ulcers from one-eighth to one-third inch in diameter.

Fully three-fourths of the right lung is hepatized and contains a few necrotic caseous masses. The pleura of this lung covered with considerable pale yellowish exudate and adherent to pericardium. Left lung and pleura normal.

No. 4. July 22. Animal died quite unexpectedly last night. Slight exudative peritonitis. Spleen engorged.

Digestive tract. Stomach normal, filled with food. Mucosa of large intestine extensively pigmented, merging into hyperæmia in the lower colon. Numerous elevated button-like ulcers one-fourth to three-fourths inch in diameter, extending from cæcum to lower colon.

In both lungs a small number of pneumonic foci, apparently of an acute character. Slight quantity of pleuritic exudate in the right pleural sac.

Lymphatic glands in general with cortex engorged with blood. Petechiæ in the cortex of kidneys, beneath peritoneal covering of diaphragm, stomach, and walls of abdomen.

No. 5. July 22. Pig died between 10 a. m. and 6 p. m.; examined at 7 p. m. Superficial lymphatic glands of the groin hyperæmic throughout; cortex has a purplish color. In the digestive tract the mucosa of the fundus of the stomach involved in hemorrhagic inflammation with superficial necrosis. In the cæcum and upper colon the mucosa is superficially necrosed.

In both pleural sacs a considerable amount of pale yellow serum. The interlobular tissue of the lungs distended with it. No hepatization. Fatty degeneration of the cortex of kidneys.

This was the first animal from which cultures were attempted. A tube of gelatine was inoculated with a bit of spleen tissue. Subsequently liquefaction took place. The turbid liquefied mass covered by a viscid pellicle. The small bacillus causing this liquefaction grew in the same manner in bouillon. It formed a viscid surface membrane very speedily, leaving the liquid clear.

An agar tube which was inoculated with a bit of spleen tissue remained sterile. Similarly two agar tubes inoculated with a loop of serum from both pleural cavities failed to develop. A pipette of pleural effusion collected at the autopsy and sealed was examined subsequently at the laboratory. It contained a large spore-bearing bacillus and some short rods in chains. In order to detect the presence of any swine-plague germs in the serum, a bouillon culture was also made from the serum, in which of course the several germs developed together. A rabbit inoculated in the ear with a drop of this culture remained well. Neither swine-plague nor hog-cholera bacteria were thus obtained from this animal.

No. 6. Pig killed in what appears to be a dying condition. No lesions found in lungs or digestive tract. Spleen somewhat enlarged but firm, not congested.

From the spleen of this animal a minute bit was placed in tube of gelatine and of agar. Both remained permanently free from growth.

From the liver two similar cultures were made. In the gelatine tube a fungus had developed after a week, but no bacteria. In the agar tube large spore-bearing bacilli appeared in the condensation water.

Neither swine-plague nor hog-cholera bacilli were obtained from this case.

No. 7. White pig, died in course of the day. Examined with Nos. 5 and 6. Ecchymoses on adductor muscles of the thigh. Hemorrhagic condition of cortex of lymphatic glands. Petechiæ under serosa of small and large intestines. Hemorrhagic inflammation of serous covering of ureters and bladder. Cortex of kidneys enlarged, pale, fatty. Lungs and digestive tract not affected.

A bit of spleen placed in a tube of gelatine and a tube of agar failed to give rise to any growth in these tubes subsequently.

In an agar tube inoculated with a bit of liver tissue, a large spore-bearing bacillus was found which grew only in the condensation water and not on the surface.

No. 8. Pig killed in a dying condition. Affected with umbilical hernia. The mucosa of the large intestine was deeply pigmented, both lungs extensively hepatized. Spleen moderately enlarged. A gelatine and an agar tube inoculated with bits of spleen tissue, both negative.

No. 9. Died in course of the day; examined between 7 and 8 p. m. Spleen large, gorged with blood. Pneumonia in localized regions throughout both lungs. Lung-worms present. The mucosa of the large intestine of a dark bluish color with scattering whitish spots of slight exudation. From this case only an agar tube was inoculated with a particle of spleen tissue. On July 24, two days later, there had appeared on the inclined surface of the agar about thirty isolated colonies, circular, slightly convex, 1½ millimetres in diameter, grayish, translucent. The clear condensation water contained a flocculent deposit. Colonies and condensation water show minute oval cocci apparently identical with swine-plague bacteria.

From these colonies a peptone bouillon tube, an agar tube, and a gelatine tube were inoculated August 25. On the following day the liquid culture contained minute clumps adhering to sides of tube. Liquid nearly clear. The agar tube showed a delicate grayish line on the surface. Both tubes contained only the minute oval cocci. The gelatine tube remained *free from growth.* To test the pathogenic nature of the germ a rabbit was inoculated in the ear from a colony of the original spleen culture July 28. (This rabbit had been inoculated June 3 into the thigh with a very attenuated culture of swine-plague bacteria sent to the laboratory.) The rabbit died in 2½ days. From the original inoculation an abscess as large as a hen's egg had formed on the superficial muscular layer of the thigh. Spleen much enlarged, soft. In stained cover-glass preparations of spleen, liver, and blood numerous swine-plague bacteria were detected. A tube of agar was inoculated from the blood and a tube of bouillon from the spleen. On the following day a uniform grayish glistening layer had appeared on the agar surface, made up, so far as the microscopic examination could go, of non-motile oval cocci. The bouillon culture, uniformly clouded, contained the same bacteria only. These were readily identified as swine-plague bacteria.

No. 10. This animal died during the night; examined at 7 a. m. next morning, July 23. Subcutaneous lymphatic glands in the groin very large, in part hemorrhagic. Spleen enlarged, firm, not congested. In the digestive tract the mucosa of cæcum is concealed by a very thin necrotic layer; mucosa itself purplish. This congestion extends along entire colon ; meso-colic lymphatics hemorrhagic.

In thorax, the base of all lobes of both lungs involved in bright red hepatization. Pleural sacs contain considerable serum which distends also the interlobular tissue. Heart muscle quite pale and discolored in patches.

A small number of tube cultures in agar and gelatine were prepared at the autopsy as follows :

An agar tube inoculated with a bit of spleen tissue showed no growth on the following day. Condensation water turbid, however, containing a large spore-bearing bacillus, the bacillus becoming spindle-shaped or nearly spherical when the spore fully matured. No other bacteria detected. On the second day a faint growth had extended on the agar surface, and a small motile bacillus detected. Agar plates were then made to isolate this bacillus, which was quite easily accomplished since the large spore-bearing bacillus failed to grow on plates. The small bacillus formed grayish, slightly gelatinous surface colonies. It was actively motile, the motion being chiefly a twirl with little change of place. Careful subsequent tests showed it to be entirely different from the hog-cholera bacillus with which it might have been confounded.

From the kidney two minute bits of tissue were transferred, one to a tube of agar, the other to gelatine. The agar tube contained on the following day about fifty colonies of the same germ found in the spleen culture of No. 9, i. e., swine plague. In the condensation water the large spore-bearing bacillus was also present, and a few gas bubbles in the agar. The gelatine tube showed very slight liquefaction of the gelatine on the surface where the bit of kidney tissue lay. This was probably due to the large bacillus which did not develop any more in this situation.

From the liver an agar and a gelatine tube culture were prepared in the same manner.

In the agar tube a considerable number of colonies of the swine-plague bacteria had appeared July 25. No other germs subsequently detected. In the gelatine culture slight liquefaction took place, owing to the presence of an actively motile bacillus in chains.

An agar tube inoculated from the pleural effusion remained sterile.

With the swine-plague bacteria from the liver a rabbit was inoculated in the ear by pricking the skin with lancet and inserting a platinum wire dipped into a colony on agar. Rabbit dead next morning. Inoculation wound on ear bluish and all veins distended. Similarly veins of subcutis filled with blood. Spleen enlarged, congested. Liver in large part invaded by coccidia. In the blood and spleen numerous bacteria detected showing the polar stain. An agar and a bouillon culture from the blood and spleen contain subsequently only the same germs found in the tissues. Bonillon faintly clouded, no clumps present. Agar growth like that from case 9.

Sections from the lung tissue of case 10 hardened in alcohol were stained in alkaline methylene blue and alum carmine. The interlobular tissue was greatly distended, containing a network of fibrin and very few cells. The alveoli contained a slight amount of fibrin and very few cells. Throughout the specimens were individual filaments and bundles of filaments of a rather large bacillus, evidently the same as that obtained in one culture. Occasionally in the contents of the alveoli were seen scattered bacteria of the size and form of swine-plague bacteria.

Sections from a subcutaneous inguinal gland hardened in alcohol and stained in alkaline methylene blue revealed large areas infiltrated with blood corpuscles and penetrated by distended blood vessels. In addition to the large bacilli found in the lung tissue clumps of very small oval bacteria, identical in appearance with swine-plague bacteria, were found scattered over the entire section, the clumps being largest in the hemorrhagic area.

In sections from the spleen none but the large bacilli were detected. The spleen was extensively loaded with blood pigment.

The two following cases were examined by Dr. Kilborne on the farm on the morning of July 28. Both had succumbed during the night. No bacteriological examination was made, but the notes are given to show the extreme variation in the lesions of this outbreak.

No. 11. Subcutaneous inguinal glands very large, hemorrhagic on section. Spleen very large, dark, friable. Mucosa of fundus of stomach very hyperæmic. Numerous petechiæ in mucosa of small intestine. Mucosa of large intestine intensely congested, portions almost hemorrhagic; depressed ulcers on Peyer's patch near valve, covered with a thin slough; one is over one-half inch in diameter. In the upper colon are large branny patches of superficial necrosis, besides numerous smaller, round points of necrosis. The various abdominal lymph glands with very hyperæmic cortex.

In thorax both pleural sacs contain 100 cubic centimetres of clouded serum. The costal and pulmonary pleura covered with a yellowish, friable, membranous exudate, causing mutual adhesion of the various lobes. Pericardial sac not invaded. The major portion of both lungs hepatized, only the principal lobes being in part still pervious. Hepatized lobes dark red. Bronchial glands hemorrhagic.

No. 12. Peritonitis indicated by the presence of 100 to 150 cubic centimetres of serous exudate in abdominal cavity. Some few fibrils of coagulated lymph stretched over intestines; extensive serous effusion between layers of meso-colon. Spleen affected like that of No. 11, though less markedly so.

Stomach contains about a litre of food; mucosa pale and covered with abundant tenacious mucus. Slight ecchymosis of upper duodenum. In lower duodenum and some feet in length of jejunum the walls infiltrated with extravasated blood to twice the normal thickness. Mucosa of a deep red color and partly covered by patches of friable yellowish exudate readily removable. Considerable blood in lower small intestine with slight ecchymosis of mucosa.

Mucosa of large intestine more or less pigmented. In cæcum are three ulcers one-fourth to one-half inch in diameter, the subjacent tissue being thickened by infiltration from one-half to three-fourths inch. On Peyer's patch near valve, and in upper and middle colon, are about half a dozen similar large depressed ulcers with very thick base, and some smaller ones.

In the lungs the cephalic and ventral lobes are involved in pale red hepatization without pleurisy.

The abdominal and thoracic lymph glands more or less hemorrhagic throughout their substance.

The following five sick animals were sent by express to the Experiment Station of the Bureau July 28. They reached the Station next day and were placed in a disinfected pen.

No. 13 dies July 31, 3 p. m.; placed in refrigerator until August 1. Weighs about 30 pounds. Skin on ventral aspect of body more or less reddened; over the sternum a few excoriations. The enlarged inguinal glands show as lumps under the skin. On section they present a mottled gray and red surface, the red limited chiefly to the cortex. Œdema of the subcutis over right knee.

False membrane covers the left half of the mass of intestines and the spleen ; consists of an elastic, rather firm yellowish white layer. Spleen firmly glued to the surrounding intestines, slightly enlarged, dark, softened. Liver firm, cuts with considerable difficulty. Kidneys in condition of parenchymatous degeneration. One hemorrhagic spot in medullary portion of one kidney. Pelvis contains a whitish glairy liquid.

Digestive tract. Two superficial necroses on the inner surface of lower lip in front, one on the upper lip and on edge of tongue near tip. Stomach contains a little deeply bile-stained fluid. Mucosa sprinkled with red spots of a washed-out appearance, most numerous in fundus and near pyloric valve. Hyperæmia of duodenum begins sharply at pyloric valve. From the opening of bile duct a few drops of thick bile can be expressed. Remainder of small intestine not markedly changed. The Peyer's patch in lower ileum has some of its follicles enlarged from which caseous masses can be expressed.

Large intestine contains much sand and gravel. Mucosa of cæcum of a dark slate color. The summit of the folds of a purplish hue. Free edge of valve bordered by a thin slough. On Peyer's patch near valve areas of necrotic tissue of a yellowish color, resting on a firm, yellowish-white base three-sixteenths of an inch thick. Upper colon has its mucosa of the same dark slate color, merging into a wine red. Two ulcers one-eighth of an inch in diameter observed. In lower colon congestion slight and gradually disappearing towards rectum. A small number of circular whitish erosions, apparently associated with the solitary follicles.

Thorax. Of the left lung, the ventral and cephalic lobes are interspersed with small regions of collapse. The remainder of the lobes very emphysematous and hyperæmic. Of the right lung, the anterior half (i. e., including cephalic, ventral, and adjacent portion of principal lobe) hepatized, covered by a thin false membrane, gluing the various lobes lightly to each other and to chest wall. The diseased lobes show the regular mottling in the upper, dorsal portion. As we proceed towards the ventral portion the mottling is less distinct, the tissue firmer and interspersed with small, irregular, necrotic foci. The smaller bronchi contain a thick, whitish pus. In ventral lobe a portion of the parenchyma as large as a marble completely converted into a grayish-yellow homogeneous mass. Of the principal lobe about one-third or one-fourth hepatized. The mottling of surface very regular. On section grayish, circumscribed areas one-half inch in diameter interspersed. Over these masses the pleura is converted into a wrinkled, roughened, hide-like membrane.

Trachea and bronchi contain small quantities of foamy liquid intermingled with yellowish particles. Bronchial glands barely enlarged, firm; some lobules pale, others reddened.

Bacteriological notes. At the autopsy an agar tube was inoculated with a platinum loop lightly rubbed over the pleural exudate. On following day a thin grayish growth with condensation water clouded. Examination of hanging drop and stained coverglass preparations shows only swine-plague germs.

In cover-glass preparations of hepatized lung tissue a large number of germs resembling swine-plague bacteria were seen, also some other forms. Two agar plates were made from the lung by thoroughly scorching the pleural surface, cutting out of the scorched area a bit of tissue and transferring with platinum loop the serum collecting in this minute cavity to 10 cubic centimetres sterilized water. With one loop of water 10 cubic centimetres agar was inoculated for Plate A. From this agar two loops were transferred to agar for Plate B. On following day about five colonies appeared on Plate A. These were made up of various bacteria. Plate B remained sterile. At the same time a bouillon culture was made from the lung tissue. In it several germs grew. These were plated. No swine-plague germs obtained.

With a bit of hepatized lung tissue, obtained as described above, a rabbit was inoculated subcutaneously August 12, 2 p. m. The rabbit was dead next morning at 8 a. m., i. e., in less than 18 hours. In spleen, liver, and blood preparations numerous polar-stained swine-plague germs present. An agar culture from heart's blood contained only swine-plague germs.

From the peritoneal exudate of pig, consisting of cells and fibrin and numerous bacteria of several varieties, two agar plate cultures were made. On Plate A one large colony of spore-bearing bacilli and one small colony of swine-plague germs. Plate B completely overgrown by the spore-bearing bacillus.

A bouillon culture from the exudate contains streptococci and swine-plague bacteria. Agar plate cultures were made therefrom and both germs isolated.

At the same time a large rabbit was inoculated subcutaneously with a bit of the exudate. Dead within 18 hours. Stained cover-glass preparations of blood from heart, spleen, and liver tissue show polar-stained swine-plague germs. An agar culture from the blood contains only swine-plague germs.

From the spleen, after thoroughly scorching through the exudate, two agar plate cultures and a bouillon culture were made with bits of spleen pulp. The bouillon culture contained only swine-plague germs. On Plate A several hundred colonies appeared next day. The superficial ones from 2 to 3 millimetres in diameter with circular, sharply outlined, slightly convex disk of a grayish glistening appearance. Those examined were made up of swine-plague germs. Plate B remained free from growth.

The swine-plague bacteria were thus obtained from lungs, pleura, spleen, and peritoneal exudate, while hog-cholera bacilli, which were looked for with the greatest care, could not be found.

No. 14. Died late August 3, and was kept in refrigerator until next morning. Before death frequent, liquid, yellowish discharges were observed.

On the left side, extending from sternum to the left forelimb, the subcutis over the pectoral muscle is infiltrated with a gelatinous blood-stained serum. The right pectoral muscle somewhat discolored by extension of the process. A discolored, blood-stained area in the abdominal muscle on the left side.

Considerable post-mortem changes, in spite of the fact that the animal was kept on ice over night. Spleen very large, dark, soft; the tissue surrounding it stained with coloring matter of the blood.

Mucosa of stomach in fundus very hyperaemic, partly hemorrhagic. The cardiac expansion covered with a thin, blackish, removable layer. In the large intestine the lymphoid patch near valve is partly necrosed. About 6 inches below valve a firm neoplastic button, about 1 inch in diameter and one-fourth inch thick. In caecum and parts of colon are numerous small (one-eighth inch) superficial necroses. Mucosa injected.

Lungs considerably diseased. Of the right lung the ventral, adjacent portion of principal, and small median lobe are hepatized. On the left side both cephalic and ventral are solidified. These diseased lobes are several times larger than the normal collapsed lung, very firm and nodular to the touch. They are covered by shreds and patches of exudate, gluing the lobes firmly to each other, to the pericardium and diaphragm. The nodular condition is due to the presence of a large number of firm, yellowish, caseous masses, varying from one-fourth to three-fourths inch in diameter, dispersed through the hepatized tissue.

Bacteriological examination included the spleen, the lungs, and the pleural cavity. At the autopsy an agar tube was inoculated with a bit of pleural exudate. Small colonies appeared on the agar surface on the following day, some made up of what appeared to be streptococci, others of micrococci. From this a bouillon culture was prepared, and the latter plated on the following day. From these plates a coccus, growing in clumps and short chains, and a large bacillus, were isolated. At the same time a rabbit was inoculated in the ear from the agar culture. The rabbit died in 11 days. At the point of inoculation was a large ulcer. Spleen small. Cultures from various organs on agar and in bouillon remained sterile.

With a bit of hepatized lung tissue a rabbit was inoculated in the ear. Dead in 8 days. Ulcer at the point of inoculation. Right lung hepatized. Pleural cavity contains some blood-stained serum. Cultures from this animal likewise remain sterile.

From a bit of lung tissue agar plates were also made. On plate A only two colonies appeared made up of large micrococci ; on plate B a thin grayish growth made up of spore-bearing bacilli.

From the spleen two agar plates were made, from which a large coccus and *bacillus fluorescens* were isolated. A bouillon culture made directly from the spleen was also plated with the result of finding a streptococcus and a large micrococcus.

No. 15. Male pig, weight about 35 pounds. Died yesterday and at once placed in refrigerator until morning.

On abdomen and inner aspect of thighs a few reddish scabs. Spleen somewhat enlarged and congested.

Digestive tract : One ulcer on tip of tongue. Stomach contains a small quantity of muddy liquid. Cardiac expansion of œsophagus covered by a yellowish, easily removable layer of friable material. The mucosa dotted with small red pits. In fundus a deeply reddened area of small extent covered with a thin necrotic layer. Duodenum, commencing with pyloric valve, of a slate color and deeply pigmented in spots. Pigmentation and aborescent injection extends down the small intestine. In large intestine considerable pigmentation of mucosa. In cæcum and upper 12 inches of colon are a large number of extensive ulcers of irregular outline, varying in length from one-half to several inches. They are slightly depressed and covered by ochre-yellow sloughs scraped away with difficulty. The entire depth of mucosa necrosed. Ileocæcal valve completely encircled by a band of necrosis. Below the first 12 inches of colon the necrosed areas are slightly raised above the surface. No marked thickening or infiltration beneath them. In addition to the larger patches there are small, slightly depressed, round ulcers one-eighth to three-sixteenths inch in diameter, with adherent superficial slough.

Of the lung tissue, a portion of the right ventral, an adjacent portion of principal lobe, and part of left ventral lobe collapsed, of a red flesh color ; no pneumonic infiltration perceptible. In the bronchi and extending into branches are small quantities of translucent, very gelatinous mucus. In the collapsed right ventral lobe the small air tubes contain whitish cylindrical plugs of mucus and pus.

Urine very turbid, contains much calcic oxalate but no albumen. Bacteriological examination was limited to the lungs, spleen, liver, and kidneys.

From the collapsed lung tissue agar-plate cultures were made with a minute bit of tissue. Plate A on following day contained numerous isolated and one spreading colony; the latter made up of motile spore-bearing bacilli, the former of swineplague bacteria. Plate B showed but two grayish flat colonies, composed of large cocci.

A rabbit inoculated subcutaneously with a bit of lung tissue was found dead on the morning of the second day. At the point of inoculation more or less extravasation of blood. About 25 cubic centimetres of blood-stained serum in abdominal cavity. Liver has a red clay color. Cultures on agar and bouillon from heart's blood and spleen contain only swine-plague bacteria.

Cover-glass preparations of spleen pulp from pig showed some large (post-mortem) bacilli. On an agar plate both swine-plague bacteria and motile bacilli resembling hog cholera were isolated. These motile bacilli were carefully studied and compared with hog-cholera bacilli. The results are given farther on.*

From the liver, in which a few large bacilli were observed in cover-glass preparations, a bouillon and an agar culture were prepared. In both there appeared swine-plague bacteria and streptococci. These were isolated on agar plates. A rabbit inoculated from the bouillon culture by a prick in the ear was dead in less than 20 hours. On small intestine numerous ecchymoses. Spleen somewhat enlarged; lungs hyperæmic. From blood and spleen cultures were made and cover-glass preparations examined. All showed swine-plague germs and these only.

From the kidney, which apparently contained no bacteria, an agar and a bouillon culture were made with minute bits of tissue. The latter remained sterile. The former showed flaky masses of micrococci resembling the swine-plague germ but had no effect on a rabbit inoculated therewith. To test virulence of the swine-plague bacteria isolated on agar plates from lung and spleen, bouillon cultures inoculated from colonies on the plates were injected into two rabbits subcutaneously, each receiving one-eighth cubic centimetre. Both were dead next day. In the organs were swine-plague germs in large numbers. The same germs obtained in culture from these rabbits.

Gelatine roll cultures made from these bouillon cultures failed as usual to develop.

No. 16. Female pig, weight from 75 to 80 lbs. Died at noon, August 4; examined 2 hours later. General condition of body good. On ventral surface along median line and on left fore-limb, slightly elevated scabs one-half inch in diameter. Extensive serous infiltration of superficial muscular tissue of ventral aspect of neck and sternum and of muscles under scapula, extending along the muscles of the left limb as far as the toes. The limb is very much swollen, the skin bluish red. Subcutaneous lymph glands of groin enlarged, firm, and pale, mottled with red lines and streaks.

Spleen enlarged, dark, and soft. The fat of the mesentery dotted with petechiæ.

Digestive tract: One ulcer on left margin of tongue. The fundus of stomach has its mucosa thrown into small folds and deeply reddened over an area of about 8 inches in diameter, the hyperæmia being most intense on the periphery of this area. About 3 inches from the pyloric valve an area of necrosed tissue 3 inches long and 1 inch wide, the slough blackish, involving the entire mucous layer and one-eighth to one-fourth inch thick. Lymphatic glands on lesser curvature with hemorrhagic cortex. Mucosa of duodenum deeply pigmented, the pigmentation beginning abruptly at the valve and extending with variable intensity throughout the entire small intestine. Mesenteric glands with cortex much reddened.

Contents of large intestine of an earthy character, more or less adherent to the mucous membrane. Mucosa of caecum and colon of a pale slate color. In the rectum a considerable number of somewhat faded punctiform extravasations on a pale mucosa. In caecum, 2 inches from valve, one ulcer, one-half inch in diameter, with superficial, pultaceous slough. On Peyer's patch near valve one small ulcer. In the upper colon about 3 inches from valve, an irregular patch of necrosed mucosa, about 2 inches long, with blackish surface and rather firm, yellowish base one-eighth to one-fourth inch thick. In remainder of colon three superficial healing ulcers with easily detachable slough.

Liver enlarged; interlobular marking appears broadened and pale, the center of acini dark red. Parenchyma somewhat softened and hyperæmic. No sensation of grittiness imparted to knife. Much fat observed in sections of fresh tissue in the form of large globules. Cells sharply outlined, nucleus distinct.

Kidneys show under capsule a small number of petechiæ. Organ very pale throughout, especially cortex.

*See page 77.

Over both lungs the interlobular tissue is distended with clear, yellowish serum, some in pleural sacs. Parenchyma œdematous, not hepatized, with exception of a necrotic and œdematous focus near tip of right principal lobe adjacent to large bronchus, which at this point is occluded with adult lung worms. Trachea and bronchi contain much foam and yellowish flakes. Bronchial glands enlarged; some lobules with cortex hemorrhagic, others mottled red and pale.

Heart cavities and valves normal. Left auricle sprinkled thickly with punctiform hemorrhages.

Bacteriological examination: Two agar tubes inoculated with a loop of pleural serum remained permanently clear.

Agar plates from the œdematous lung developed on plate A about 150 colonies of *bacillus fluorescens*. On plate B one similar colony. A rabbit inoculated with a bit of the same material remained unaffected.

Spleen pulp shows no bacteria on microscopic examination. One agar plate prepared with a bit of spleen pulp shows growth only around the bit of tissue. The growth made up of motile bacilli which resemble hog-cholera bacilli. The notes of the comparative studies of these bacilli are reserved for the end of this chapter.* .

A tube of bouillon inoculated from the spleen remained clear.

Liver tissue shows no bacteria on microscopic examination. Two agar plates prepared with a bit of tissue remain sterile. A tube of bouillon inoculated in the same manner became clouded with a short, thick, non-motile bacillus. Agar plates from one kidney likewise negative.

Pig No. 17 died August 5, in the morning, and was examined soon after. Female, weighing about 40 pounds, and in good condition.

The subcutaneous lymph glands in the groin are enlarged and reddened, in part hemorrhagic. In peritoneal cavity, a considerable quantity of straw-colored serum. The serosa of large intestine roughened and covered by shreds of exudate. A considerable mass of gelatinous exudate between layers of mesocolon. Mesocolic glands swollen, very hyperæmic. Spleen very large, dark colored, friable (11 by 3 by ¾ inch).

Digestive tract: Stomach partly filled with a thick pea-soup like liquid. Fundus over an area of 4 inches in diameter, deeply congested and swollen. A few ascarides in stomach. The mucosa of ileum sprinkled with numerous punctiform hemorrhages. The cæcum appears as an enormously enlarged, dark, bluish-gray body, the color being due to diffuse blood extravasations. To it are adherent several coils of small intestine and of colon. It is impossible to separate these parts, the adhesion being due to extensive inflammatory deposits. The wall of cæcum shows great thickening; in several places it is 1½ inches thick. The mucosa of cæcum and upper colon sprinkled with large numbers of ochre-yellow excrescences about one-fourth inch apart, and from one-eighth to one-half inch in diameter. These are readily pulled away from the membrane, some leaving but a faint depression, others a roughened spot behind, surrounded by a bluish-red zone. The larger the excrescences the deeper the necrosis beneath them. The mucosa is in general of a dirty slate color, faintest in the rectum.

Lungs normal with following exceptions: An area of collapse 1 inch square in right cephalic lobe, and a small area on diaphragmatic surface of right principal lobe about one-half inch square showing consolidation beneath. Bronchial glands and those at base of heart with cortex hemorrhagic. Air tubes free from mucus and lung worms.

A considerable quantity of serum in heart-sac. Epicardium is covered with a very delicate exudate and is thickened and roughened. Ecchymoses on left auricle and endocardium of left ventricle.

Liver not markedly changed; bile very thick and flaky. Parenchymatous degeneration of cortex of kidneys. Some petechiæ on pyramids and in pelvis.

Bacteriological examination: From the pericardial exudate one agar and one bouillon tube inoculated with platinum loop. Both remained sterile. Similarly two tubes inoculated with peritoneal exudate remained sterile. Agar plate cultures from the spleen which showed no germs on microscopic examination were made by adding to tube A a bit of spleen pulp, to B three loops from first tube. Plate A developed one colony made up of micrococci. On plate B one colony made up of bacilli appeared. A bouillon culture made at the same time with bit of spleen pulp became clouded and contained a large spore-bearing bacillus and a micrococcus. This was plated, but only one colony developed made up of cocci. A rabbit inoculated at same time from bouillon remained unaffected. From the liver agar plates remained sterile.

The important pathological changes found in these 17 cases varied greatly from animal to animal. In some cases the lungs were most severely involved, in others the lesions in the large intestine must be regarded as most important. In this respect this outbreak differs very decidedly from the two preceding ones, in which there was more or less uniformity observable. Besides the respiratory and digestive tracts, the lymphatic glands and the serous membranes were most frequently involved. In the glands tumefaction with hyperæmia and hemorrhage was quite common. In the lungs the appearance of the hepatized regions varied considerably. In some necrotic changes had already appeared. The following synopsis of the lesions will serve to illustrate the statements made:

No. 1. July 21, 1890. Pneumonia of left lung with exudative pleuritis. Hyperæmia of stomach; diphtheritic inflammation of large intestine. Lymphatic glands hyperæmic and hemorrhagic. Spleen very large.

No. 2. Double serous pleuritis. Large intestine hyperæmic and pigmented. Several ulcers.

No. 3. Pneumonia of right lung with necrotic masses; pleuritis, pericarditis. Ulceration of upper large intestine.

No. 4. Pneumonic foci in both lungs; pleuritis. Hyperæmia, pigmentation, and ulceration of large intestine. More or less blood extravasation in lymphatics.

No. 5. July 22. Double serous pleuritis with interlobular œdema of lungs. No hepatization. Hemorrhagic inflammation of stomach; similar inflammation of large intestine with superficial necrosis of mucosa.

No. 6. Killed; examination negative.

No. 7. Hemorrhagic lymphatics; subserous hemorrhages in abdomen. Lungs and digestive tract not affected.

No. 8. Extensive hepatization of both lungs; pigmentation of large intestine.

No. 9. Spleen large. Scattering pneumonic foci; lung worms. Hyperæmia of large intestine.

No. 10. Partial hepatization of lungs with interlobular œdema and serous pleuritis. Lymphatics hemorrhagic. Hyperæmia and superficial necrosis in large intestine.

No. 11. July 28. Extensive hepatization of lungs with double plastic pleuritis. Hyperæmia of stomach. Diphtheritic inflammation of large intestine. Lymph glands hyperæmic and hemorrhagic.

No. 12. Anterior small lobes of lungs hepatized; no pleuritis. Peritonitis. Hemorrhagic inflammation of a portion of small intestine. Pigmentation of large intestine; indurated ulcers present. Lymph glands hemorrhagic.

No. 13. July 31. Partial hepatization of lungs with necrotic foci; pleuritis; bronchitis; peritonitis. Hyperæmia of large intestine; ulceration slight.

No. 14. August 4. Gelatinous œdema of left fore limb. One-half of lung tissue hepatized with many necrotic foci; pleuritis. Hemorrhagic inflammation of stomach. Ulcers in large intestine

No. 15. Atelectasis of several lobes of lungs; no hepatization; bronchitis. Hemorrhagic inflammation of stomach. Extensive ulceration in large intestine.

No. 16. Extensive œdema of left fore limb. (Edema of lungs. Serum in both pleural sacs. Bronchitis. Lung worms. General swelling of lymph glands. Hyperæmia of stomach with localized necrosis. Pigmentation and discoloration of small and large intestines. A few ulcers in the latter.

No. 17. August 5. Lungs nearly normal. Pericarditis. Peritonitis. Hyperæmia of stomach. Enormous inflammatory thickening of cæcum with deposits around it. Spleen large. Lymph glands swollen, hyperæmic.

The important question arises as to the true nature of this disease. To the writer it appeared at first like hog cholera, possibly like a mixed outbreak of hog cholera and swine plague. The bacteriological investigations, however, did not confirm this opinion, based on the postmortem appearances. As the investigation proceeded hog-cholera bacilli failed to appear in the cultures with certain exceptions to be discussed farther on, and indicated in the following table as motile bacilli:

Case No.	Lungs.	Pleura.	Spleen.	Liver.	Kidney.
5	....	Negative	Negative	....	....
6	....	....	...do	Negative	....
7	....	....	... do	....do	....
8	....	....	....do	....	....
9	....	....	Swine plague.	....	....
10	....	Negative	Negative	Swine plague.	Swine plague.
13	S w i n e plague.	Swine plague	Swine-plague.	(Peritoneum) swine plague.	....
14	Negative .	Negative	Negative	....	....
15	S w i n e plague.	....	Swine-plague and motile bacilli.	Swine plague .	Negative .....
16	Negative .	Negative	Motile bacilli (h o g cholera ?)	Negative	....do ....
17	....	(Pericardium) negative.	Negative	....do	(Peritoneum) negative.

It will be noted that in the eleven cases bacteriologically examined swine-plague bacteria appeared in only four. In these four cases they were detected in nearly all organs subjected to examination. Why they were not found in all cases may be due to several reasons. In the first place, former experience has shown that swine-plague bacteria are apt to remain localized, and that dissemination through the body does not always take place. If limited to the digestive tract they could not have been detected, because this was not subjected to examination. Secondly, swine-plague bacteria are short lived, even in cultures. They may have largely disappeared from the body at the time of death. Moreover, it is not improbable that many swine die from the secondary effects of the disease. (See No. 17.) The time of examination is therefore of importance. Swine may be infected all together within a short time, and the retarded deaths may be due to partial resistance followed by complications. The lesions produced in the lungs and intestines may permit other bacteria to enter the body, which complicate still more our understanding of the real cause.

While therefore the bacteriological results were meager, the inoculation experiments and one experiment in which pigs were exposed to the disease were successful in demonstrating conclusively the pathogenic power of the swine-plague bacteria obtained from this outbreak.

EXPOSURE EXPERIMENT.

The experiment in which healthy station pigs were exposed to those sent from Pleasantville in the same pen is particularly interesting. Five Station pigs (Nos. 383–387) were placed in the pen with the five infected pigs (Nos. 13–17). The period of exposure varied greatly, as may be seen from the table below. Nos. 384 and 385 received a thorough exposure so far as regards contact with the sick pigs, No. 383 a partial one, and Nos. 386 and 387 were placed in the infected pen only after all the sick and infected had died.

No.	July 29.	July 30.	July 31.	Aug. 1.	Aug. 2.	Aug. 3.	Aug. 4.	Aug. 5.	Bacteriological result.
13	×	×	Dies						Swine-plague bacteria.
14	×	×	×	×	×	×	Dies		Negative.
15	×	×	×	×	×	Dies			Swine-plague bacteria.
16	×	×	×	×	×	×	Dies		Negative (hog-cholera bacteria?)
17	×	×	×	×	×	×	×	Dies	Negative.
383						Exposed	×	×	
384	Exposed		×	×	×	×	×	×	
385	....do		×	×	×	×	×	×	Dies August 9 of swine plague.
386									Exposed Aug. 9.
387									Do.

The result of this exposure was the death of No. 385 of swine plague. In the lungs were disseminated numerous hepatized foci undergoing necrosis. There was considerable plastic pleuritis matting the lobes together. In the digestive tract extensive hyperæmia bordering on hemorrhage. Swine-plague bacteria were obtained from the lungs, pleural exudate, heart's blood, and large intestine. Cultures were also made from the spleen, liver, mesenteric glands, and kidneys, to make sure that if hog-cholera bacteria were present they should not be overlooked, but none could be found. A clearer demonstration of the disease-producing power of swine-plague bacteria could not well be obtained. The importance of this case warrants the publication of the notes in detail:

No. 385. Black Essex grade, aged 3 months, placed in pen containing Nos. 13 to 17, inclusive, on July 29. Found dead August 9, rather unexpectedly, after a sickness lasting but 2 days, and manifested by dullness and refusal to eat.

Examined a few hours after death; kept on ice in the meantime. As the skin is reflected, the cut subcutaneous blood vessels discharge drops of dark blood. In abdominal cavity nothing abnormal. Pericardium thickened, opaque, vessels injected. Left half of heart firmly contracted. Ecchymoses on the right auricle. Right side contains a pale clot imbedded in a dark coagulum.

Respiratory tract : Mucosa of larynx and epiglottis congested. Mucosa of trachea covered with a very thin layer of translucent, very viscid mucus. Bronchial glands enlarged, pinkish on section. Costal pleura of the right side discolored, thickened, overlaid by a soft grayish exudate ; its blood-vessels injected. Diaphragm similarly affected. The various lobes of the lungs lightly glued to one another and to the pericardium. Considerable pleuritic exudate on the right ventral and along the edges of both principal lobes. Lungs in general hyperæmic. Left cephalic lobe emphysematous ; left ventral shows hemorrhagic spots near its tip and contains two firm nodules, appearing as yellowish spots under pleura. Left principal lobe contains four of these spots corresponding to firm nodules in the parenchyma. Besides these, a wedge-shaped, very firm, hepatized mass extends inward from the edge and almost through the depth of the parenchyma. These various masses appear yellowish, homogeneous, imbedded in hyperæmic, air-containing tissue. They vary from one-eighth to five-sixteenths inch in diameter, nearly all of them situated near the surface. The largest ones are covered by roughened, thickened pleura, thrown into wrinkles ; these are in part yellowish, in part bright red and pink in color. The three lobes of the right lung contain these necrotic masses. There are several in the right cephalic, about six in the ventral, and over a dozen in the principal lobe, the largest being one-half inch in diameter.

Digestive tract : Several small superficial sloughs on dorsum of tongue, near tip. Stomach contains a considerable quantity of food. Fundus deeply reddened over an area 5 inches in diameter. Duodenum with its mucosa bluish-gray, pigmented. Peyer's patch showing as an aggregation of small, depressed pigment spots. Arborescent injection of jejunum merging into a general hyperæmia lower down. About 18 inches above valve a patch of mucosa 2 inches long, intensely reddened.

Extensive pigmentation of mucosa of cæcum. Mucous glands at the valve distended with plugs. About 12 inches below the valve the mucosa is intensely reddened, merging on hemorrhage. A very delicate elastic membrane (fibrin ?) covers this region, extending for about 12 inches down the colon. Below this latter point the mucosa continues more or less hyperæmic and pigmented into the rectum. Mesenteric glands enlarged with cortex and interlobular portions hemorrhagic. Mesocolic glands in the same condition.

Kidneys with cortex pale, somewhat enlarged. Otherwise no marked changes observed.

Spleen slightly enlarged and softened.

Bacteriological examination includes the following organs :

In two necrotic foci of the lungs examined a number of minute oval bacteria with polar stain were observed in cover-glass preparations. Plates prepared with a bit of this tissue remained sterile, however. A rabbit inoculated subcutaneously with a bit of necrosed lung tissue at 3 p. m., Aug. 9, was found dead next morning. In the spleen and heart's blood and in cultures therefrom only swine-plague bacteria detected. The cultures were inclined agar and bouillon.

In cover-glass preparations of pleural exudate, a large number of swine-plague bacteria were detected. Two inclined agar and one bouillon tube inoculated at the autopsy. One agar remained sterile, the two other tubes developed into pure cultures of swine-plague germs.

From the spleen agar plates prepared with a bit of pulp remained sterile. A bouillon culture contained a large spore-bearing bacillus. From the liver, the same preparations made. Both plates and bouillon remained sterile.

From the contents of right ventricle two agar and one bouillon tube were inoculated after scorching through the wall. The bouillon tube remained clear. Both agar tubes contained subsequently a grayish glistening growth of swine plague bacteria.

From a mesenteric gland one agar and one bouillon tube inoculated by scorching the surface and removing with small scissors bits from beneath this area. Both tubes remained sterile.

From one kidney two agar plates and a bouillon tube prepared. Plate A was spoilt by an extensive surface growth. On Plate B about 12 colonies developed, some made up of micrococci, some of bacilli. The bacilli on plate B were examined more closely and readily differentiated from hog-cholera bacilli (*bacillus coli communis*) in gelatine rolls and subcultures therefrom. The bouillon culture contained a large spore-bearing bacillus.

From the hyperæmic mucosa of colon agar plates were made, which, however, gave no information as to the presence of swine-plague bacteria. A rabbit was inoculated by injecting under the skin a few drops of scraping from the colon stirred up in sterile water. Rabbit found dead on the second morning. In cover-glass preparations and cultures from blood and spleen only swine-plague bacteria detected.

The four remaining pigs did not die, but they were ill for a time and remained in a very unthrifty condition. They were killed about three months after exposure. No pathological changes were observed with the exception of slight hyperæmia in the duodenum and a few small areas of collapse in the lobes of the right lung of 384, and bronchitis and collapse of the entire right cephalic lobe in 386. In 387 the solitary follicles of the colon were swollen; from some a curdy mass could be expressed through a central depressed opening.

INOCULATION OF SWINE WITH SWINE-PLAGUE BACTERIA FROM THIS OUTBREAK.

Besides the successful production of disease by simple contact with diseased animals, a considerable number of inoculations upon pigs were made, the earlier ones to test the pathogenic power of these bacteria, the later ones in carrying out vaccination experiments. Only a portion of them are given below:

On October 30, two pigs (Nos. 399 and 404) were fed the chopped viscera of six rabbits which had been inoculated subcutaneously on the preceding day and had all succumbed during the night. The viscera, which contained immense numbers of swine-plague germs, were readily eaten after being mixed with a small quantity of feed. The pigs remained well.

On the same day the following inoculations were made:

Pig No. 402 received a subcutaneous injection of 5½ cubic centimetres of a peptone bouillon culture of swine-plague bacteria 24 hours old, one-half injected into each thigh.

No. 400 received into a vein of the leg 1 cubic centimetre of the same culture.

No. 403 received into a vein of the leg 5 cubic centimetres of the same culture.

No. 401 received into the right lung, through the chest wall, 5 cubic centimetres.

These pigs were black pigs (Essex and Berkshire crosses) about 3 months old. The bacteria originally derived from case 15 had been passed through a number of rabbits and guinea-pigs since August 3.

No. 402 showed no signs of disease after the inoculation. No. 403

died first. The inoculation took place at 3:30 p. m., October 30, and the animal was found rigid at 6:30 next morning.

General blush of the skin on the ventral aspect of the body. On right side the small vessels of the subcutis filled with blood which oozes from the cut ends in the form of thick drops. Both superficial inguinal glands hyperæmic.

A small quantity of yellowish serum in the abdominal cavity. A network of delicate fibrils stretching across coils of intestine. Peritoneum dull, opaque in appearance.

Trachea and bronchi contain reddish froth, mucosa reddened. Punctiform hemorrhages under the pleura of left lung. Slight general œdema of both lungs; some interlobular œdema of the right.

Stomach contains a moderate quantity of deeply yellow-colored liquid. The mucosa covered with a thick layer of tenacious mucus. The fundus over an area 8 inches in diameter is intensely hyperæmic, the intensity being greatest in the center of the area. Contents of duodenum of a blackish color. More or less hyperæmia throughout the small intestine, with swelling, hyperæmia, and ecchymosis of Peyer's patches. More or less hyperæmia in patches in the cæcum and colon. The glandular patch at the valve especially reddened. Feces dry. Follicles show as circular, red spots from the serous side. Lymphatic glands of the lesser omentum, mesentery, and meso-colon very much reddened.

Liver congested. Bile thicker and darker than normal.

On surface and throughout the cortical portion of kidneys numerous punctiform hemorrhages. Pyramids intensely congested. Glands at hilus with cortex hemorrhagic. Bladder contracted, empty.

No. 401 died in the evening of October 31, i. e., somewhat more than 24 hours after inoculation. It was placed in large refrigerator until next morning.

Subcutaneous vessels as in No. 403. Small patches of a grayish, viscid exudate on the large intestine, liver, and spleen. Punctiform ecchymoses barely visible to naked eye under serosa of large intestine. Arborescent injection of the subserous vessels on ventral wall of abdomen and intestines generally.

In the right pleural sac a considerable quantity of blood-stained liquid and shreds of exudate stretching from lung to chest wall. The various lobes of both lungs glued together and to pericardium. The lateral, ventral border of both lungs have the pleural covering much thickened, roughened, and thrown into wrinkles. Both lungs congested and œdematous. In the right cephalic lobe a mass of dark red hepatized tissue which may be the place where needle punctured. Grayish-red hepatization of the major portion of left cephalic lobe. Trachea and bronchi contain a large quantity of reddish foam. Mucosa with minute vessels injected.

Pericardium opaque and roughened. The entire epicardium similarly affected. A grayish membranous exudate about the base of the heart extending upon the large vessels.

Stomach contains a moderate quantity of turbid, saffron-colored liquid; mucosa covered with a thin layer of mucus. Punctiform reddening of fundus. Small intestine contains much yellow liquid, mucosa not altered. Large intestine contains firm feces. No lesions observed.

Cortex of kidneys rather pale, pyramids very dark red.

Under capsule of spleen, some small extravasations. Parenchyma pale, organ not enlarged.

Cover-glass preparations from peritoneum of small and large intestines and of liver show many swine-plague bacteria. In the spleen their presence not determinable on cover-glass preparations. Cultures contained them, however.

No. 400 became very sick on the day following the inoculation. It was unable to rise and showed signs of distress when disturbed. It continued to lie quietly until November 3, when it died at noon, about 4 days after the inoculation:

Animal in very good condition; abundance of subcutaneous fat. Blood vessels of subcutis as in preceding cases. In abdominal cavity nothing abnormal.

Double pleuritis with adhesion of various lobes of lungs to each other, to chest wall and pericardium. Grayish membranous exudate over the ribs on the right side. The subjacent pleura has a bluish appearance. The corresponding pleura of lung covered by patches of similar exudate. The right lung dark red throughout but not hepatized. In the left lung the ventral third of the ventral lobe is dark red, solid; on section granular and interspersed with grayish areas. The lateral edge of adjacent principal lobe is likewise hepatized for a distance of 1 inch inward and 2 inches along border.

Pericardium thickened, opaque, roughened; blood vessels injected; entire epicardium covered with a rather thick false membrane, loosely attaching the pericardium to it. Large quantity of dark, partly coagulated blood in right heart; a very little in the left heart.

Stomach empty, lined with a thin layer of bile-stained mucus. In the lower ileum several patches of mucosa of a dark bluish color, each about 6 inches long. Large intestine filled with dry, firm feces. In the upper colon, mucosa is slate-colored, lower down normal in color.

Liver rather firm, surface appears slightly roughened and mottled, owing to congestion of individual lobules. These deeply reddened lobules disseminated through whole parenchyma. Hepatic vessels contain much thick dark blood.

Kidneys considerably enlarged, the surface beset with a large number of grayish, slightly elevated spots, some surrounded by a dark-red zone. They vary from one-half to one-fourth inch in diameter and are about one-fourth inch apart. On section they correspond to grayish wedge-shaped masses extending inward through the cortex and in the form of longitudinal striæ through the pyramids to the papillæ. These infarcts are made up, examined fresh, chiefly of pus corpuscles.

Only the spleen of this case was subjected to examination. No germs were seen in cover-glass preparations, but they were obtained in cultures.

Sections were prepared from the hepatized lung tissue hardened in alcohol and stained with alum-carmine and methylene blue. The alveoli were found completely occluded with cellular masses partaking chiefly of the character of leucocytes. The red blood corpuscles were present in small numbers. Fibrin not detected. The alveolar capillaries distended with red corpuscles projecting into the lumen of the alveoli. In the cell masses swine-plague bacteria are very abundant. They are scattered between the cells, not in clumps.

Sections from the kidney hardened in alcohol and stained with methylene blue showed the cellular infiltration in the cortex and extending through the medullary portion in the form of cylindrical cell masses occupying the lumen of the large collecting tubules which are completely stripped of their epithelium. Swine-plague bacteria are disseminated through these cell masses as in the lung tissue already described.

The foregoing notes indicate that 5 cubic centimetres of a peptone bouillon culture of swine-plague bacteria injected into the circulation or the lungs may be fatal within 24 hours. They also show a tendency to inflammation of the large serous cavities, especially the pleural, even when the injection is made into the circulation. This inflammation corresponds closely with that found in the disease as it occurs in nature.

Another important fact is the production of pneumonia not only when the bacteria are injected into the lungs, but when introduced into the circulation.

Feeding and subcutaneous inoculation failed to produce disease. That the latter will now and then produce disease is well shown by the following case:

Pig No. 454 was inoculated February 28, 1891, with a peptone bouillon culture of the swine-plague germ from Case 15. Three cubic centimetres were injected subcutaneously into each thigh. The culture was one day old, prepared from an agar culture. (Three others inoculated at the same time remained well.) The pig was found dead March 2, in the early morning. It had thus lived between 36 and 48 hours after the inoculation.

Black and red female pig, weighing about 50 pounds; in good condition. Considerable reddening of the skin of ventral aspect of body and ears. Subcutaneous fat reddened. From the cut vessels dark, thick blood exudes. Over both inoculated thighs the subcutaneous vessels are extensively injected, forming a dense network. The subcutis has a glistening appearance. On the right thigh, near Poupart's ligament, the subcutaneous connective tissue is thickened, yellowish opaque, and friable, over an area of several square inches.

On opening abdomen the intestines appear very much reddened. A few elastic fibrils stretched across coils and about 10 cubic centimetres of turbid serum present.

Stomach contains a large quantity of food. Mucosa not affected. Considerable catarrhal inflammation of the duodenum, which extends into jejunum. The hyperæmia extends through the small intestine, but much less intense. Several ascarides present. The mesenteric glands somewhat congested; in two of them old, caseous masses, occupying nearly the entire gland.

Large intestine contains a large quantity of semiliquid feces. Mucosa normal. Meso-colic glands hyperæmic.

In each pleural sac about 10 cubic centimetres of turbid effusion. Lungs hyperæmic and œdematous. Interlobular tissue of anterior (or cephalic) lobes slightly thickened and opaque, the result of some former inflammation. In trachea and bronchi some reddish frothy liquid. The mucosa shows marked injection of the minute blood-vessels, in some places almost hemorrhagic. In right bronchus a small number of large lung worms.

On left auricle of heart ecchymoses. In left ventricle a small quantity of rather thick dark blood. Right ventricle distended with it. Coagulation feeble.

Liver quite pale generally, acini distinctly outlined. Parenchymatous inflammation. On the surface may be seen with a hand-lens numerous minute grayish-yellow dots, one or more in a lobule and situated chiefly on the periphery. In sections of fresh tissue they appear as irregular, opaque, amorphous patches. In stained sections from tissue hardened in alcohol they appear as intralobular aggregations of round cells occupying the place of parenchyma cells. These foci are probably result of some former disease. In the intralobular capillaries occasional masses of swine-plague bacteria detected. (See Pl. xi, Fig. 4.)

Spleen enlarged and hyperæmic. Kidneys with cortex broadened. The base of pyramids somewhat darker red than normal.

In cover-glass preparations from blood and spleen, swine-plague bacteria present, in the blood in considerable numbers.

From a bit of spleen pulp two agar plates prepared. On the first, after 24 hours, numerous swine-plague colonies appeared; on the second nothing developed.

Agar cultures from peritoneal and left pleural cavity remained clear.

An agar tube inoculated from the blood contained a large number of confluent colonies. A bouillon tube became faintly clouded. Both contained only swine-plague bacteria.

An agar culture from the liver inoculated with platinum wire contained on the following day a considerable number of swine-plague colonies.

Two of the following cases are of importance owing to the peculiar lesions of the joints caused by the inoculation.

November 11. With a peptone bouillon culture of swine-plague bacteria from the same stock culture (Case 15) the following inoculations were made: Nos. 405 and 408 received into a vein of one hind leg one-half cubic centimetre; Nos. 406 and 407 received one-fourth cubic centimetre. In each case the quantity was diluted with sterile bouillon so that 1 cubic centimetre of liquid was injected.

For the two following days all pigs were sick and refused to eat the food set before them. Within a week Nos. 406 and 408 had nearly recovered, No. 405 partially, while No. 407 was unable to get up and continued so until it was killed, December 3, by a blow on the head. Its chief lesions were a peculiar necrotic and suppurative condition of the joints and suppurative pericarditis.

Condition of animal very poor. Enlargement of the left carpal and phalangeal joints, of the right elbow joint, and both hock joints. Along the inner aspect of the left scapula, large masses of yellowish dry pus deposited around the muscles. Carpal joint of the same limb much enlarged. On opening the joint a considerable quantity of dry pus is found deposited around the bones and between the tendons over the joint. The joint surfaces are discolored, the cartilages in part detached, and the small bones readily crushed. The toes enlarged, the enlargement due to similar suppurative changes around the phalanges.

Right elbow joint enlarged; joint cartilages greenish, opaque. Around the joint several abscesses containing either a turbid liquid or dry cheesy pus, which has burrowed along the intermuscular septa of the forearm. Carpal and metacarpal joints of this limb not affected.

On both hind limbs the tarsal and metatarsal joints very much enlarged and in the same condition as those just described. In a small number of ribs the sternal articulation involved in suppuration. Over the lower ribs on the right side the intermuscular septa are imbedded in masses of dry cheesy pus.

Within the lymphatic glands at the angle of the jaw and in the inguinal region are minute yellowish masses.

Lungs normal. Pericardium thickened, adherent to the heart surface by means of a layer of brittle, straw-colored pus, covering the entire heart surface with exception of a small area on the left ventricle, to which the pericardium is attached by means of delicate fibrous tissue. The pus is most abundant at the base.

The digestive tract free from inflammatory changes. Stomach contains a small quantity of liquid of a deep yellow color, and some mucus.

Kidneys show on the surface a small number of discolored, slightly depressed spots, corresponding to pale, whitish, wedge-shaped infarcts, extending into medullary portion.

Liver discolored in spots and patches on the surface. These also seen in sections. Gall bladder contains a dark-greenish solid mass, cutting like firm cheese and filling up the entire space of the bladder.

One tarsal joint saved and opened for bacteriological examination. The skin is removed, the sac of the abscess thoroughly scorched and opened through the scorched area with a flamed knife. A considerable quantity of greenish-yellow liquid holding in suspension brittle masses of pus wells out. From this liquid one inclined agar, one bouillon tube, and three agar plates prepared. The agar and the bouillon tube contained active growths of the injected swine-plague germ. On the first agar plate an immense number of minute colonies appeared; on the second about two hundred, and on the third very few. From these plates all colonies examined were made up of swine-plague bacteria.

A rabbit inoculated subcutaneously with a particle of the pus died within 20 hours. In cover-glass preparations from liver and spleen, and in an agar culture from the latter organ, swine-plague germs demonstrated.

Another case of suppurative changes in the joints following intravenous inoculation is the following:

No. 411 received, December 4, 1¼ cubic centimetres bouillon culture of swine-plague bacteria into a vein of one hind limb. It immediately became very sick, was unable to get up, and died December 22. It was greatly emaciated. Both carpal and the left tarsal joint enlarged, the changes within the joint similar to those described in the preceding case. Several rib joints in the same condition. In this case also the presence of the virulent swine-plague bacteria in one diseased carpal joint was demonstrated by inoculating two rabbits with pus therefrom. Both succumbed within 20 hours to the inoculation. In the organs the characteristic polar-stained bacteria. An agar and a peptone bouillon culture from the spleen of the pig remained sterile.

Two other pigs inoculated at the same time and with the same dose died in 15 hours with commencing peritonitis, pleuritis, and pulmonary œdema. In both the stomach was hyperæmic.

PATHOGENIC EFFECT ON SMALLER ANIMALS.

The great virulence of the swine-plague bacteria from this outbreak, as compared with those of former outbreaks, is even better shown by the inoculation of small animals. In the notes given below it will be seen that not only rabbits but guinea-pigs and pigeons succumb to very small subcutaneous doses of the growth from cultures, while large fowls are killed by inoculations into the muscular tissue. The inoculation of guinea-pigs and pigeons with cultures from previous outbreaks was usually uncertain even though rabbits invariably succumbed.

Guinea-pigs.—October 2, 1890. One guinea-pig received subcutaneously into the thigh one-fourth cubic centimetre of a peptone-bouillon culture of swine-plague bacteria, a second one-twelfth cubic centimetre, and a third one twenty-fourth cubic centimetre. The liquid in each case was diluted with 2 parts of sterile bouillon, hence 3 times this quantity of liquid was actually injected.

The second guinea-pig died within 24 hours. At the point of inoculation considerable gelatinous œdema, the blood vessels injected, and the muscles of thigh covered with a thin, grayish layer. Stomach and small intestine hyperæmic. Spleen enlarged, dark colored. In blood, spleen, and liver very few bacteria. An agar culture from the spleen developed only colonies of swine-plague bacteria. The first guinea-pig died several hours later with similar lesions. Peyer's patches hyperæmic. A moderate number of swine-plague bacteria in the various organs. The third guinea-pig died in about 36 hours with the same lesions.

October 7. A guinea-pig was inoculated with one one-hundredth cubic centimetre of a peptone-bouillon culture. It died in 40 hours. The small intestine very hyperæmic, occasional patches of punctiform hemorrhages in mucosa. A small number of bacteria in the various organs not showing a distinct polar stain.

October 14. Two guinea-pigs were inoculated, one with one one-thousandth cubic centimetre and the other with one forty-thousandth cubic centimetre of a peptone-bouillon culture. The first died in 36 to 40 hours, the second in 8 days. In this latter case there was considerable purulent thickening of the subcutis around the place of injection. The intestines, liver, and spleen covered by a thin layer of cellular and fibrinous exudate. Very few bacteria in the various organs and the exudate.

Numerous additional inoculations of very small doses into guinea-pigs confirmed the fatal effect of this organism on this species of animals.

Mice.—November 15, 1890. Two gray mice inoculated with one or two drops of peptone-bouillon culture died within 20 hours. In the spleen numerous swine-plague bacteria which show the polar stain very well in stained preparations.

Pigeons.—No. 1 inoculated by an injection of 0.3 cubic centimetre of a turbid suspension from an agar culture 24 hours old. The injection was made under the skin of one pectoral muscle. Pigeon No. 2 inoculated in the same way with 0.2 cubic centimetre, the needle of the syringe penetrating superficially the pectoral muscle.

No. 2 died next morning. Parboiled appearance of inoculated muscle. Liver remarkably pale and firm. In blood and liver immense numbers of bacteria showing polar stain very beautifully.

No. 1 dies in 24 hours. Slight subcutaneous infiltration at point of inoculation. Liver in same condition as in No. 2. Considerable hyperæmia of mucosa of duodenum and adjoining small intestine. Contents stained reddish. Bacteria not so numerous as in preceding case.

These inoculations having proved successful, two pigeons were inoculated with very small doses from an agar culture from pigeon No. 2, three days old.

No. 3. Skin over pectoral incised with a lancet and a loop dipped into the culture and rubbed into subcutis.

No. 4. Skin incised and a minute portion on platinum wire rubbed into subcutis. Both dead next morning, i. e., within 18 hours. Bacteria abundant as in preceding cases. Livers pale and firm.

Fowls.—Two adult hens, inoculated in the same manner as pigeons Nos. 3 and 4, but with a slightly increased quantity on the platinum loop, remained unaffected.

Two other adult hens were inoculated with the same bacteria from an agar culture suspended in bouillon until turbid. A hypodermic syringe was used and the needle passed superficially into the fibers of the pectoral muscle. One fowl received one-fourth cubic centimetre, the other one-half cubic centimetre.

The first died in 24 hours. No local reaction perceptible. The liver is very pale; sprinkled over it many whitish points, character not determinable. Intestines normal except cæca, which have some ecchymoses under serosa. Large number of bacteria in blood and liver showing polar staining very well.

The second fowl died in 36 hours. Liver very pale and sprinkled with ecchymoses along the course of the superficial vessels. In liver few, in blood large numbers of the injected bacteria.

TWO BACILLI ISOLATED FROM THIS OUTBREAK CLOSELY RESEMBLING HOG-CHOLERA BACILLI.

The facts thus far given are sufficient to prove that in this outbreak the swine-plague bacteria were the chief if not the only cause of the epizoötic. In the course of the investigations two kinds of bacteria were obtained, which deserve special attention owing to their resemblance to hog-cholera bacilli.

Owing to preoccupation of the writer with Texas fever investigations the cultures from cases 13 to 17, inclusive, were made largely by Dr. V. A. Moore, assistant in the laboratory, and he was directed to pay special attention to all motile bacteria having any resemblance to hog-cholera bacilli. Four of such were isolated, one from the lung of No. 14, one from the spleen of No. 15, one from the spleen of No. 16, and one from the kidney of No. 385. These four bacilli the writer subjected to a very careful examination. Those from Nos. 14 and 385 were found to be identical with the common intestinal germ, *bacillus coli communis*. Those from the spleen of Nos. 15 and 16 grew very much like hog-cholera bacilli on gelatine plates and in rolls. To bring out the differences observed we will call the bacilli from Case 15 *γ* and those from Case 16 *δ*.

In gelatine rolls *δ* could not be distinguished from a parallel roll culture of virulent hog-cholera bacilli. The surface colonies of *γ* presented a somewhat different type, in that they spread in a thinner layer with very thin edges, somewhat like *bacillus coli*. They differed, however, from colonies of the latter by their very restricted growth, attaining a diameter of only 1½ to 2 millimetres.

Both caused considerable turbidity of peptone bouillon, while virulent hog-cholera bacilli cause only cloudiness, which very rarely merges into a moderate turbidity. These bacilli multiplied therefore far more energetically in bouillon than do virulent hog-cholera bacilli. There were other slight differences between these two bacilli. Thus, while *γ* caused uniform turbidity of the culture fluid, *δ* grew more or less in clumps, which caused a rapid settling of the growth in spite of the motility of the bacilli. When the culture was shaken up numerous clumps and flakes rose from the bottom. In the hanging drop these bacilli presented a strange appearance. The various clumps, composed of 10 or more bacilli, moved rapidly in various directions across the field of the microscope. This peculiarity of *δ* maintained itself after passing through several rabbits and many cultures. In their pathogenic power these two bacilli differed not only from hog-cholera bacilli, as they are usually encountered in outbreaks, but from one another, as the following experiments show :

Bacillus *γ*. Plate cultures made from original culture and bouillon inoculated from a colony. When 24 hours old, this culture was used to inoculate 2 rather large rabbits.

One white rabbit received subcutaneously 0.3 cubic centimetre culture liquid.

One black rabbit received into an ear-vein 0.3 cubic centimetre culture liquid.

Two weeks thereafter, neither having shown any signs of disease, they were reinoculated. The black one received 1 cubic centimetre, the white 0.5 cubic centimetre, both into an ear vein.

The white rabbit remained unaffected. It was killed after 16 days, but no lesions were found. The black one died in 36 hours. The blood was thick, tarry; the left lung hypostatic; the spleen small. No bacteria in cover-glass preparations from the spleen. This result did not place these bacilli above the level of the ordinary intestinal bacteria (*bacillus coli*): for 1 cubic centimetre of the latter produces death in rabbits with equal promptness when injected into an ear vein. Two additional intravenous inoculations were made April 28:

One rabbit, weighing about 3 pounds, 0.3 cubic centimetre bouillon culture.

Two rabbits, weighing about 4 pounds, 0.6 cubic centimetre bouillon culture.

Both rabbits remained apparently unaffected.

Finally, on May 21, two rabbits, weighing each 3 pounds, were inoculated by receiving into an ear vein 0.6 cubic centimetre of a bouillon culture. The agar culture had been replated, and from a colony this bouillon culture was prepared. Neither rabbit showed signs of disease.

It did not seem worth while to spend any more time on this obviously non pathogenic organism. Its marked resemblance in morphological and biological characters to the hog-cholera bacillus makes it highly probable that it is closely related to this latter germ.

Bacillus δ. Similar experiments with this bacillus from case 16 proved that this one did possess pathogenic properties, though feeble in character.

From the original culture a gelatine plate was prepared and a peptone bouillon tube inoculated from a colony. When 4 days old two small yellow rabbits were inoculated. One received about 0.4 cubic centimetre under the skin, the other the same quantity into the ear vein. The latter rabbit died in 6 days.

The spleen is very large, dark, and softened, and contains a large number of what appear to be hog-cholera bacilli. The liver shows areas of necrosis. The gall-bladder attached by inflammatory exudation to the omentum. The lungs œdematous. Fatty degeneration of cortex of kidneys. The small intestine filled with a glairy yellowish liquid. The cultures from this case contained only the injected bacilli, exhibiting in bouillon tubes the peculiar characters described above.

The rabbit which received the subcutaneous dose remained apparently unaffected. After 22 days it was inoculated with swine plague and found dead the following morning. At the place of the first inoculation a cyst nearly 1 inch in diameter containing curdy pus. In the appendix vermiformis from twenty to thirty whitish nodules under serosa. Similar nodules on Peyer's patch near ileo-cæcal valve.

A rabbit which received an injection of 0.3 cubic centimetre into abdomen remained well. When killed, 10 days after the inoculation, the same appearances of appendix of cæcum were found as those just described.

Several weeks later, with the culture obtained from the first rabbit, four rabbits were inoculated, each by an injection into an ear vein of 0.2 cubic centimetre. These all died within 36 to 48 hours. In the one which lived 48 hours the following lesions were observed :

Large, dark spleen. Interlobular network of a yellowish color through the whole liver. Under the microscope this appeared as fatty degeneration of the periphery of the acini. Lungs œdematous. Left ventricle of heart in advanced fatty degeneration. Cloudy swelling of kidneys. In the spleen large numbers of bacilli which appear somewhat larger than hog-cholera bacilli and are usually in pairs. Cultures from this case contain only the injected bacilli.

That the bacillus δ is an attenuated variety of the hog-cholera bacillus can not be doubted when these rabbit inoculations are taken into consideration. There was, moreover, a gradual increase in virulence observed from one inoculation to the subsequent one. Thus the first rabbit, which was a small one, received 0.4 cubic centimetre bouillon culture into a vein, and lived 6 days. The second rabbit received 0.2 cubic centimetre, and died in 2 days. Subsequent inoculations showed that .05 cubic centimetre was fatal in a few days, while smaller quantities caused a disease from which the rabbits recovered.

We have thus seen that in addition to the swine-plague bacteria, which produced disease both after inoculation and exposure to diseased swine, two motile bacteria were isolated from cases 15 and 16, but from no other. These both resembled hog-cholera bacilli very closely, and undoubtedly are related to them. But they differed from each other in several particulars, notably in virulence. While the one from No. 16 was shown to be an undoubted but very attenuated variety of hog-cholera bacilli, the one from No. 15, though it may also be an attenuated variety, has so little virulence that its bearing upon the outbreak in question must be regarded as entirely negative. As to the other bacillus the question is not so simple. It may be seriously doubted whether it had anything to do in producing the disease, since its presence was detected in but one out of eleven cases, and since hog-cholera bacilli from genuine hog-cholera outbreaks appear quite regularly in cultures from the spleen. Moreover, the virulence of the latter is many times greater than that of the bacilli in question. Thus, to test this matter thoroughly, one pig received into the lungs 5 cubic centimetres of a bouillon culture, another 5 cubic centimetres into the abdomen, and several others 7 cubic centimetres each directly into the blood. None of them showed any signs of illness after such treatment. Lastly, two pigs were fed, after a fast of 24 hours, with 200 cubic centimetres of a bouillon culture each. This feeding was repeated on two successive days. A slight diarrhea, lasting a few days, was the only visible effect.

On the other hand it is not to be denied that these bacilli may have occasioned a part at least of the intestinal disease observed in this outbreak. The truth may be that these attenuated hog-cholera bacilli, brought from distant outbreaks by individual animals which have survived such outbreaks or proved themselves insusceptible at the time, had no effect on the animals brought in contact with them until the swine plague broke out, when they may have started into activity and contributed to the fatality of the disease. That the bacillus δ from case 16 should have been the cause of the outbreak and the swine-plague bacteria secondary to it would seem very far-fetched in the light furnished by the experiments with the swine-plague bacteria.

In addition to the attenuated hog-cholera bacilli, another disease germ was isolated from cultures prepared on the farm from some of the earlier cases. This germ is briefly referred to in the autopsy notes as a rather large bacillus, which multiplied in the condensation water of agar cultures but not on the surface. It produces spores which are formed in the central portions of the rod. The latter during the development of the spore becomes spindle-shaped. This bacillus does not multiply in culture tubes under ordinary conditions, but requires media more or less free from oxygen. It belongs to the group of anaërobic bacteria, and is probably identical with the bacillus of malignant œdema. Whether it is the bacillus which I have frequently detected in swine and called "post-mortem" bacillus it is impossible to state, since I have made no special effort to cultivate the latter and test its virulence on animals.

The bacillus in question grows in deep layers of agar in test tubes. The isolated colonies develop quite slowly and after 1 or 2 weeks attain a maximum diameter of 2 millimetres. The surface of the fully developed spherical colony is closely beset with radiating finger-like projections, which are perhaps one-fourth as long as the diameter of the entire colony. An addition of glucose to the agar favors the growth of the bacilli and causes the production of considerable gas, which may break up the agar column and force portions nearly out of the tube. This bacillus also multiplies in peptone bouillon containing, say, 2 per cent. of glucose, provided oxygen is absent. For this purpose the fermentation tube, which I have found very useful in bacteriological work, is very well adapted.* When such tubes are inoculated a faint cloudiness of

* The fermentation tube has been in use for many years in various kinds of biological work. Its value in bacteriological work as a culture tube I have already called attention to (*Centralblatt f. Bakteriologie* (1890) VII, p. 502), but reproduce here a few suggestions as to its use. The tube, reduced one-half, is shown in the accompanying cut. After it has been plugged with cotton wool and sterilized in the dry hot air oven it is filled with the culture liquid and steamed on three successive days. The air collecting at the top near *b*, which has been forced out of the liquid, should be removed immediately after each steaming by gently tilting the tube. After the third steaming, the liquid in *b* is entirely freed of dissolved air. The tube is inoculated with platinum wire, loop, or pipette as are ordinary tubes. A large number of bacteria, especially those living in the intestines of animals, are capable of setting up a fermentation when glucose and other sugars are present. The gas collecting at the top in *b* is composed chiefly of CO_2 and H.

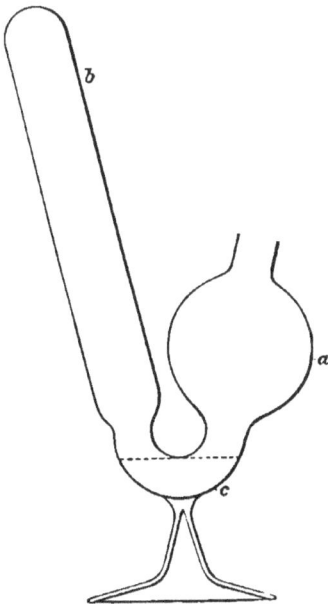

I have found the fermentation tube very valuable in the diagnosis of hog-cholera bacilli, which are capable of fermenting glucose, etc. Hog-cholera bacilli from a large number of outbreaks tested thus far all manifest this phenomenon, and it may be safe to assert that any bacilli resembling hog cholera which do not produce gas are not hog-cholera bacilli.

the liquid appears within 1 or 2 days, and gas bubbles rise in the closed branch. After a few more days the liquid in the closed branch is in part replaced by gas, which continues to form for a week or longer. Meanwhile the bacilli have become deposited in the bottom of the tube as a whitish, flocculent mass. By cultivation in this way I was able to keep these bacilli alive for many months until opportunity offered to test them on animals. That they are pathogenic and may cause malignant œdema the following inoculations are sufficient to demonstrate:

May 22, 1891, 3 p. m. From a culture in a fermentation tube 14 days old one-half cubic centimetre of the turbid deposit withdrawn and injected under the skin of a rather large guinea-pig in the region of the abdomen.

The animal, apparently well at 9 a. m. next morning, lay down at 10.a. m., and died at 1 p. m. On examination very extensive sanguinolent œdema of the subcutis over abdomen and part of thorax. On left thigh the œdema accompanied with much distension of the subcutis with gas. The serosa of abdominal cavity discolored and vessels injected. Both sides of heart contain dark soft clots. In the blood-stained subcutis large numbers of the injected bacilli, in the spleen a few, in the blood none observed. Two fermentation tubes were inoculated with blood and a particle of spleen pulp, respectively. In both a typical growth appeared in a few days.

At the same time a colony of these bacilli was removed from a glucose agar tube, now 22 days old, and placed into the subcutis of abdomen of another guinea-pig through an incision. No symptoms observed until 48 hours, when the animal rested with abdomen on floor of cage and did not stir when aroused. Apparently neither drowsy nor in pain. It was found dead on the morning of the third day (66 hours). The lesions as well as the distribution of the bacilli and cultures from blood and spleen were the same as in preceding case.

A pig inoculated subcutaneously with 2 cubic centimetres of turbid deposit from a fermentation tube showed no signs of disease.

Whether these bacilli can be implicated in the death of some cases of this outbreak in which swine-plague bacilli were not detected, these inoculations will not permit us to decide. The bacilli at the date of inoculation had been cultivated for 10 months, and hence may have become attenuated in the meantime.

X.

At the end of 1890, Veterinarian E. C. Schroeder was directed by the chief of the Bureau to make some examinations of swine diseases in the West with the object of still further determining the distribution of hog cholera and swine plague.

Among the several herds examined only one deserves mention, because positive bacteriological results were obtained. This herd was found about 1½ miles south of Chillicothe, Missouri, where greater or smaller losses from infectious swine diseases are said to occur each year.

Cultures were made by Dr. Schroeder on agar, and these were carefully examined by the writer subsequently. The cultures from one animal contained only swine-plague bacteria, those from another only hog-cholera bacilli. This outcome again illustrates the caution which must be exercised in isolating the bacteria obtained from diseased animals, so that the characters of two different bacteria may not be confounded and regarded as one, as has probably been done by Billings in his investigations. No conclusions are drawn from the meager details below as to which bacteria may have been the predominating cause of the disease. They simply illustrate the wide diffusion and intermingling of two pathogenic bacteria.

The following brief synopsis has been condensed from Dr. Schroeder's notes of the post-mortem examination:

No. 1. December 27, 1890. Dead several days. Animal had been opened by owner. Spleen greatly congested. Intestines could not be examined. Right lung almost completely hepatized. Epicarditis. Lymph glands in general much congested.

No. 2. January 2, 1891. Several foci of hepatization in lungs. Muco-purulent condition of smaller air tubes. Spleen very large, dark, softened. Petechiæ in cortical portion of kidneys. Much mucus in stomach. Feces in large intestine very dry, coated with mucus. Patches of hyperæmia; no ulceration.

No. 3. January 3. On pubic region a large open wound, with subjacent tissue very much thickened by inflammatory deposits. Spleen very large, dark, and friable. Slight exudative peritonitis. Congestion of fundus of stomach. Feces in large intestine exceedingly dry and covered with mucus. Several inches below valve a blackish, necrotic patch, 1½ inches long and a half-inch wide (Peyer's patch?). Mucosa congested in isolated patches. No other ulceration observed. Lungs hyperæmic. No hepatization. Hemorrhagic condition of auricles of heart. Petechiæ on epicardium of ventricles.

No. 4. January 3. Black female, killed by a blow on head. Spleen as in No. 3. Digestive tract normal, with exception of patches of congestion in large intestine and what appear to be two small ulcers on ileo-cæcal valve (enlarged mouths of glands?). Slight exudative peritonitis. Lungs free from hepatization.

Bacteriological examination. From No. 1 no cultures made, owing to post-mortem changes.

From No. 2 bits of the spleen placed in two agar tubes. On January 19 there were about 100 isolated and confluent colonies on the agar surface in one of the tubes. In the other no growth had appeared. These colonies were carefully examined and found to be swine-plague bacteria. Many of the colonies were dead, as the agar had dried out somewhat. Transferred to bouillon and other media, the diagnosis was confirmed. These bacteria were, however, more or less attenuated as regards pathogenic effect.

January 31. Two white rabbits of medium size inoculated, one subcutaneously on one ear with a loop dipped into condensation water of an agar culture, the other under skin of abdomen in same manner. Both rabbits were very quiet for several days, the former with ear drooping and very much reddened. On the eleventh day it was chloroformed. The greater part of the inoculated ear blackish; hard, like a board; necrosed. A purulent inflammation extending from ear down on the face and neck. Heart muscle pale and flabby. Much fat in the fibers. Fatty degeneration of convoluted tubes in cortex of kidneys. Numerous granular casts in urine. Spleen small and pale. Fatty condition of liver.

The second rabbit was well at this time. When killed a small abscess in subcutis, with ecchymosis of contiguous abdominal muscles. No internal lesions.

From No. 3 agar cultures had been made from spleen and liver. Both developed, and evidently contained the same bacteria. The liver culture contained a considerable number of isolated colonies. From these gelatin plates were made. The various culture tests and the microscopic characters (size, motility, etc.), all pointed to hog-cholera bacilli. The diagnosis was confirmed by the following inoculations:

One rabbit received into an ear vein one-fifth cubic centimeter bouillon culture derived from a colony on a gelatine plate. One rabbit received same dose of same culture under skin of flank.

The first rabbit died within 48 hours. No local lesion. Spleen very large, dark, and firm. Liver fatty. Lungs slightly œdematous. In the spleen numerous bacteria, resembling hog-cholera bacilli in every way. Cultures confirmatory.

The second rabbit died in 5 days. At the place of inoculation purulent infiltration of skin and subcutis with ecchymoses on periphery. Spleen large, dark, firm. Heart muscle fatty. Lungs œdematous and hyperæmic along ventral border. In liver interlobular tissue broadened, pale, representing degeneration of periphery of acini. Cultures contain only hog cholera bacilli.

From pig No. 4 an agar culture from the spleen failed to develop.

XI.

An intermingling of hog cholera and swine plague was also observed in two outbreaks studied in 1889, and very briefly referred to in the report of the Secretary of Agriculture for the same year (p. 75). These outbreaks are of importance, in so far as the hog-cholera bacilli obtained therefrom were somewhat modified with reference to their biological and pathogenic properties. It was evident also that these hog-cholera bacilli were the predominating cause of the disease in many of the animals exposed, and the full report is therefore reserved for another publication. The investigations are here mentioned to illustrate once again the wide distribution and frequent intermingling of these two pathogenic bacteria.

THE BACTERIA OF SWINE PLAGUE.

If a cover-glass preparation from the spleen pulp of a rabbit, which has succumbed 16 to 20 hours after inoculation, be stained in alkaline methylene blue for a few minutes, and examined with a high power either in water or after it has been permanently mounted in balsam, a very large number of swine-plague bacteria will usually be found among the cells of the spleen pulp.

These bacteria at first give the impression of very minute flattened cocci in pairs, the individuals of each pair separated from each other by a small intervening space. (See Plate XI, Fig. 1.) A closer inspection, however, shows that each pair is in reality a single elongated body of which the two cocci are the stained extremities. The intervening space is the unstained connecting body, the borders of which are continuous with those of the stained extremities as indicated in the figure.

The two end pieces have usually a lunar shape, the concavities turned toward each other. The depth of the concavity varies somewhat and may even be replaced by a straight line, in which case the end pieces have a hemispherical shape.

It is highly probable that the bacteria as described above are in a state of division, the two stained extremities representing the two future cells, and the intervening space perhaps a common membrane without any contents. During this process the protoplasm retracts from the central portion of the rod and gathers at the two extremities.

In stained cover-glass preparations these bacteria are not infrequently found varying in length. This variation is mainly due to the variation in length of the middle unstained piece. Thus preparations of the same germ from some rabbits may show a very short middle piece so that the two concave borders of the comparatively large end pieces appear to touch each other on the lateral margins of the bacterium. In other preparations this middle piece may be from two to three times the size of each end piece. In a preparation from the spleen of a rabbit inoculated with the German swine plague the middle piece was in many bacteria so long as to suggest involution or degenerate forms. It made the entire rod one and one-half to two times longer than the normal forms. In some of these bacteria one end piece had divided and the double germ resulting therefrom appeared like a long, feebly stained bacillus, in which there were in both ends and in the middle of the rod the deeply stained round coccus-like bodies. In all the variations the stained extremities do not vary much in size. This is, therefore, an additional reason why these extremities should be regarded as daughter cells in process of separation from one another.

The foregoing description applies to bacteria taken directly from the dead animal and dried on cover glasses. In this dried condition after they have been stained and mounted in balsam they are about 1 μ* long and 0.5 to 0.6 μ wide. Their ends are rounded off and in general their form is that of a somewhat elongated oval. Besides these average forms there may be others, 1.8 μ long and 0.7 to 0.8 μ broad. Dimensions larger than these generally belong to forms evidently abnormal in development. In these the width generally remains the same as that of the shorter forms.

While these bacteria in cover-glass preparations from pigs, inoculated rabbits, etc., are the same in appearance, the same bacteria in sections of tissues hardened in alcohol and stained in the same way do not correspond to the description just given. They are smaller than the smallest forms described above, and as a rule do not show distinctly the unstained middle piece. They appear under high powers as minute uniformly stained oval bodies. It is evident that in the one case the drying process has a tendency to flatten objects out against the cover glass, while in the hardening process there is a tendency to shrinking, which is not counteracted in any way. Hence the different appearance may be due simply to the difference in the mode of preparation.

The foregoing description applies to virulent varieties which cause death of the inoculated rabbit in 16 to 20 hours. In the various organs and the blood the inoculated bacteria are present in large numbers. The more attenuated varieties do not produce death thus quickly. The rabbit dies in from 3 to 10 days. The bacteria have meanwhile become localized in the peritoneal cavity or the pleural cavity, and produced an inflammatory exudate which contains immense numbers of bacteria. These do not stain so well as those described, and rarely show the polar arrangement of the protoplasm distinctly. There is reason to suppose that many of these forms are already destroyed by the inflammatory process.

In cultures the swine-plague bacteria are smaller than in the bodies of inoculated animals. Examined in water they appear so minute that it is with difficulty that they are detected at all (Zeiss apochr. 2 millimetres, compens. oc. 4). In general they answer well to the description of micrococci, although they are not round but oval in form. On the border of the drop holding them in suspension the polar arrangement of the protoplasm is occasionally detected.

Swine-plague bacteria are non-motile. This important character

---

* Micromillimetre or about one twenty-five thousandth of an inch.

serves to distinguish them at once from hog-cholera bacilli. In liquids an active Brownian motion is frequently seen which is so violent at times as to give the impression of spontaneous motion. They do not change their place, however, and this agitation is thus readily distinguished from the very active flagellar movement of hog-cholera bacilli. Again, flagella are easily demonstrated on the latter. Dr. V. A. Moore has devoted much time in the laboratory to bring out flagella on swine-plague bacteria without success. This was of course to be expected. No spores have been observed in any of the conditions under which they have been cultivated. They are destroyed in liquids by a temperature of 58° C. (136.4° F.) in 7 minutes. Their rapid destruction by drying, disinfectants, etc., renders it highly probable that no spores are produced.

The question whether swine-plague bacteria should be called bacilli or micrococci is not an easy one to settle satisfactorily. In the tissues and very rarely in cultures they may appear as elongated rods, but only under abnormal conditions. Their usual form, both in cultures and in sections from tissues, is that of an oval, the longer diameter exceeding but slightly the shorter diameter. While we are inclined to call them micrococci as least misleading, usage seems to have followed the Koch school in calling them bacilli. In the present report they are simply denominated bacteria, and the matter of nomenclature is left open.

Swine-plague bacteria are not so easily cultivated as hog-cholera bacilli. Besides refusing to multiply on certain media in which hog-cholera bacilli readily grow, their life in cultures is much shorter. In the following the chief characters of these bacteria in the various culture media are given somewhat in detail, since no one character is sufficiently peculiar, and all are necessary to positively recognize the species.

The growth on nutrient gelatine is variable and therefore not reliable. The bacteria from outbreaks VII, VIII, and IX, as a rule, refused to multiply in gelatine. Occasionally colonies develop in roll cultures, but the growth is very feeble and retarded, and may be overlooked unless they are carefully watched for longer than a week. An increase of alkali seems to favor their growth.

If we turn to the bacteria discovered in 1886 and 1887 perhaps the only biological difference observable is the more abundant growth of these varieties in nutrient gelatine. The deep colonies were from one-fourth to one-half millimetre in diameter; the surface colonies were 4 or 5 times as large. The former appeared after some days with a pale margin, the central portion being brownish, granular.* The failure to grow on gelatine can not be regarded as entirely due to the lower temperature in which gelatine must be kept. It seems partly due to the nature of the medium, partly to the adaptability of the bacteria to be cultivated.

* See p. 89 and Plate IV in the report of the Bureau of Animal Industry for 1886, and Plate XII, Fig. 3, of this report.

The same method being employed in the preparation of gelatine and agar, the fact that swine-plague bacteria will develop in an agar tube when a gelatine tube by its side inoculated in precisely the same manner fails to develop shows that some physical or chemical condition of the gelatine is at fault. The gelatine must be regarded as a drier medium than the agar, which dryness is not so favorable. Another possible explanation of the refusal of the more recently cultivated varieties to multiply in gelatine may be due in part to slight improvements and modifications in the preparation of nutrient gelatine and agar. On the whole gelatine should not be used in the investigation of swine plague.

On nutrient agar prepared in the ordinary way with peptone bouillon and kept in the thermostat, multiplication goes on rapidly, so that in 24 hours the deep colonies are one-eighth to one-fourth millimetre in diameter according to the proximity of the colonies to one another (see Plate XII, Figs. 1 and 2). The deep or submerged colonies appear roundish or lenticular, and when examined under a low power (about 60 diameters) they are brownish, opaque, with margin and surface beset with small knobs, thus giving the disk a reticulated and the border an irregular wavy appearance. Some colonies do not present this appearance, but remain smooth on surface and margin. The surface colonies are glistening, slightly convex, whitish disks, barely translucent. They are circular, with margin having no irregularities. Under a low power the central portion is brownish, granular, towards the margin becoming more homogenous and translucent and exhibiting usually very delicate radial striations. The deep colonies may attain a diameter of one-half to three-fourths millimetre; the surface colonies, when 1 centimetre apart, measure 4 to 5 millimetres in diameter. As in the case of other bacteria, the size of the colonies varies inversely as their number on the plate.

A very characteristic feature of such plates is the strong, disagreeable, pungent odor which is emitted. This is best detected when the agar plates have been prepared in so-called double dishes, and these are quickly opened after they have been closed for 24 hours or less. When the growth is abundant this odor persists even after repeated exposure of the plates.

On the inclined surface of nutrient agar in tubes, the growth may appear either in the form of isolated colonies or as a single grayish translucent patch, depending on the number of bacteria rubbed upon the surface originally. (See Plate XII, Fig. 4.) The condensation water collected in the bottom is usually quite turbid within 24 hours. After two or more weeks it will be found that the growth in the water has assumed a viscid gelatinous consistency, so that it tends to come away in a mass when a platinum loop is introduced. This has been characteristic of all swine-plague germs thus far examined.

Bouillon containing from one-fourth to 1 per cent. peptone becomes uniformly but very faintly clouded within 24 hours. From some outbreaks the growth may be at first flocculent or granular. The bacteria grow in clumps, leaving the liquid unclouded. This, however, is no permanent character, since cultures of succeeding generations after a time become uniformly clouded. Not infrequently a partial membrane or a ring growth on the glass at the surface of the liquid appears after one or more weeks of quiet standing in a uniform temperature. In the bottom a deposit forms during this time which may become quite viscid. No marked changes in reaction occur in the cultures even after some weeks of multiplication. There seems, however, to be a slight tendency toward the production of an acid reaction in culture liquids originally alkaline. No fermentation of sugars accompanied by the liberation of gases takes place. In this particular these bacteria are distinguished from hog-cholera bacilli, which are able to cause fermentation of glucose with liberation of hydrogen and carbonic dioxide.*

Milk inoculated undergoes no changes visible to the naked eye. The reaction becomes faintly acid. On the surface of boiled potato there is no appreciable multiplication. I have once or twice observed a very faint whitish growth which may have been due to the culture material transferred. In general, however, swine-plague bacteria do not grow on potato.

The temperature range for the multiplication of swine-plague bacteria may be placed between 65° and 108° F. The growth is very feeble between 65° and 70° F., and most rapid and abundant between 97° to 100° F. The nature of the culture media seems to have some effect. If this is very favorable multiplication may take place at a slightly lower temperature than when less favorable.

Among the properties of the group of swine-plague bacteria pointed out by German observers, and valuable as a means of diagnosis, is the capacity to form in culture liquids indol (C_8H_7N) and phenol or carbolic acid (C_6H_6O). Both substances are also formed ordinarily during the decomposition of albumins, and hence are present in the intestinal tract.

The method followed in testing for these substances is that suggested by Lewandowski,† which consists in distilling the culture liquid, say 250 cubic centimetres of ordinary peptone bouillon in which the bacteria have grown for 10 days, with 50 cubic centimetres of strong chlorhydric acid and testing the first portions of the distillate for indol and phenol separately. For indol 2 cubic centimetres of a 25 per cent solution of sulphuric acid is added to 5 cubic centimetres of the distillate, and then 2 to 3 drops of a 0.1 per cent solution of sodium nitrite. A red coloration indicates the presence of indol. A fine crystalline precipitate forming immediately, or soon after the addition of bromine water to the distillate, indicates the presence of phenol.

---

* See foot-note, p. 81.
† Deutsche med. Wochenschrift. 1890, S. 1186.

The method followed by the writer was practically the same, excepting that potassium nitrite was used, the solution being prepared fresh each time. A preliminary trial having shown phenol in most cultures, but only a trace of indol in one culture, a second test was made.

Slightly alkaline bouillon containing 1 per cent peptone and one-half per cent salt was sterilized in flasks containing 200 cubic centimetres each, and inoculated May 15, 1891, with swine-plague bacteria from the following sources:

1. Outbreak VII.
2. Outbreak VIII.
3. Outbreak IX.
4. German swine plague (virulent variety).

These cultures were allowed to remain in the thermostat until June 8, at which time they were tested according to the method indicated.

The culture from VII had an abundant viscid growth on the surface attached to glass. Liquid turbid when shaken, faintly alkaline; odor disagreeable and characteristic of swine-plague cultures. Slight precipitate of tribromphenol; no indol reaction.

The culture from VIII had a thick membrane on the surface, somewhat viscid. Other characters as in preceding case. Phenol less abundant. No indol.

The culture from IX like that of VIII. Phenol very abundant. Very faint indol reaction.

The culture of German swine plague gave nearly as much precipitate of tribromphenol crystals as the preceding, but no indol reaction.

Various cultures of hog-cholera bacilli tested showed the presence of neither indol nor phenol.

The reason why the indol reaction failed in our hands is not clear. It may be that as the culture grows older the phenol reaction increases while the indol reaction may disappear. It will be noted that phenol was detected in all four cultures, in the third in abundance. The crystals of tribromphenol examined microscopically were identical in all four cases. It might also be mentioned that the second culture, inoculated with the least virulent swine-plague bacteria, contained the smallest quantity of phenol.

RESISTANCE OF SWINE-PLAGUE BACTERIA TO DESTRUCTIVE AGENTS.

Swine-plague bacteria possess less power to resist destructive agents than hog-cholera bacilli. Their life, even under what might be regarded as the most favorable conditions, is brief. In the laboratory cultures are liable to die out in 3 or 4 weeks, especially when evaporation is going on. In bouillon but 2 or 3 days old the large majority of bacteria are dead, for if plate cultures be made from the liquid only very few colonies appear where we would expect thousands. The resistance to drying is feebler than that of hog-cholera bacilli. In the report for 1886, some experiments are reported which show that drops from a bouillon culture dried on cover glasses failed to infect fresh tubes on and after 3 days. Shorter periods were not tried in this experiment.

In another trial with peritoneal exudate from a rabbit containing immense numbers of swine-plague bacteria dried on cover glasses, bouillon tubes were infected up to the third day, but not thereafter.

In the report for 1887–'88, p. 143, some additional experiments are given with swine plague bacteria from Iowa. Bacteria in bouillon cultures, dried for a period longer than 36 hours, were destroyed, while bacteria from agar cultures lived for 6 days. The difference is very likely due to the thickness of the dried film, which is much greater when material from agar cultures is employed, owing to the density of the growth.

They are likewise more easily destroyed by disinfectants than are hog cholera bacilli. Hence the extended experiments made in connection with our study of hog-cholera bacilli have made a repetition of such experiments with swine plague unnecessary. We simply give the following experiments with lime, since this disinfectant is cheap, efficient, and easily applied.

April 18, 1891. Five cubic centimetres of limewater is thoroughly mixed with a loop of agar growth of swine-plague bacteria 1 day old from outbreak IX. Bouillon tubes, inoculated with a loop of this suspension after 45 minutes, 1, 2, 3, and 6 hours, remain clear.

This experiment was repeated with the modification that two loops of agar growth were mixed with the limewater and bouillon tubes were inoculated immediately after making the mixture, and after 5, 10, 15, 20, 25, 30, 40, 50, and 60 minutes. All but the first tube remained clear.

The limewater in a third trial was diluted with two volumes of water, and therefore contained but 0.04 per cent of lime approximately. In this series all but the tube, inoculated immediately after the bacteria were mixed with the limewater, remained clear.

These experiments show that as weak a solution of limewater as 0.04 per cent is sufficient to destroy swine-plague bacteria almost immediately.

The rapidity with which swine-plague bacteria are destroyed by drying and other agencies made it very probable that their life in the superficial layers of the soil where they are deposited must be very brief. The following simple experiments place this supposition beyond doubt, for they show that rabbits can no longer be infected after the bacteria have been in the soil for a period of 8 to 10 days.

March 7, 1888. A pot of sterilized soil was inoculated by pouring upon its surface 50 cubic centimetres of a beef-infusion peptone culture which had been growing for 24 hours at 95° F. The pot was kept in the laboratory, covered with a disinfected bell jar. March 10 a rabbit was inoculated with one-fourth cubic centimetre of sterile beef infusion in which a little soil had been stirred up. The rabbit died in 48 hours with numerous polar-stained swine-plague bacteria in the internal organs. Ten and fourteen days respectively after the soil had been infected two other rabbits were inoculated from the same soil. Both remained well.

March 29. A pot of soil was inoculated by pouring upon its surface 100 cubic centimetres of a beef-infusion peptone culture 2 days old of swine-plague bacteria obtained through Prof. W. H. Welch from an outbreak studied by him in 1887 in Baltimore, Md. The pot was sunk into a larger pot filled with sterilized soil and the whole buried in the garden of the Department of Agriculture on a level with the soil. To test the virulence of the culture used to infect the soil a rabbit was inoculated in the ear with a lancet dipped into it. It died within 48 hours. Cover-glass preparations and cultures reveal the inoculated bacteria in the internal organs. Four days later, a rabbit was inoculated from the soil as in the preceding experiment. The rabbit died within 2 days. No bacteria found on cover-glass preparations; cultures contain several kinds of bacteria. Swine-plague bacteria not detected. This case was therefore doubtful. April 7, nine days after the infection of the soil, another rabbit was inoculated therefrom. This one remained well.

Some experiments made in 1886 indicated a rapid destruction of swine-plague bacteria in sterilized water. In 1888 and 1889 Dr. V. A. Moore was charged to carry out another series of experiments to test this problem more thoroughly.

In the first experiment a platinum loop rubbed over the surface growth of an agar culture 24 hours old was stirred up in 10 cubic centimetres sterilized Potomac water and placed in the laboratory. A second tube of water prepared in the same way was placed in the refrigerator (50°-55° F.).

From the first tube 1 cubic centimetre transferred to bouillon on the fifth day failed to infect it. From the second tube the same quantity failed to infect a bouillon tube on the seventh day.

In a second series two tubes of water were infected, each with three loops of surface growth and three drops of condensation water from an agar culture. It will be noted that in this case some nutritive substances must have been transferred in the condensation water. Bouillon tubes were infected up to the thirty-fifth day from the water in the laboratory. From that in the refrigerator a successful inoculation was made on the nineteenth day, but it failed after 38, 41, and 44 days.

A third series was tried. Each of two tubes was infected with three loops of growth from an agar culture 1 day old. From the one in the laboratory inoculation with 1 cubic centimetre failed on the ninth day; from the one in the refrigerator, after the twentieth day.

Swine-plague bacteria in water containing no nourishment may thus succumb in 7 to 10 days at ordinary temperatures (60°-70°F.). When some nutritive substances are present this period may be materially lengthened. In lower temperatures the bacteria seem to live somewhat longer. A comparison between the relative vitality of hog-cholera and swine-plague bacteria may be tabulated as follows:

	Swine plague.	Hog cholera.
Destroyed by moist heat at 58° C	7 minutes	15 minutes.
Destroyed by drying (from bouillon)	1½ to 2 days	7 to 9 days.
Destroyed by drying (from agar)	After 6 days	After 4 months.
Destroyed in water	After 10 days	After 2 to 4 months.
Destroyed in soil	After 4 days	After 2 to 3 months.

THE PATHOGENIC ACTION OF SWINE-PLAGUE BACTERIA.

EFFECT ON SMALL ANIMALS.

In order to understand best the pathogenic action of swine-plague bacteria a brief account of the results of inoculation into smaller animals will be given before the disease in swine is discussed. These inoculations being made with pure cultures the effects are exhibited in a clearer manner than in swine in which hog cholera is so frequently an associated disease.

The various inoculations into the smaller animals, such as mice, rabbits, guinea-pigs, pigeons, and fowls, are briefly described in connection with the special outbreaks from which the bacteria were obtained. In the present chapter only the general results will be brought together and broadly outlined in their bearing on swine plague.

If we take the inoculation of rabbits as a starting point we find that there are different degrees of virulence or pathogenic activity very clearly brought out. Thus the swine-plague bacteria obtained from outbreaks VII and IX, and from Germany,* were of the maximum degree of virulence so far as rabbits are concerned, inasmuch as the latter died 16 to 20 hours after inoculation with very minute doses of culture material. The blood and the various organs contained often immense numbers of the inoculated bacteria, always exhibiting the very characteristic polar stain shown in Plate XI, Fig. 1. These bacteria were thus able to produce a rapidly fatal septicæmia.

The bacteria from some of the other outbreaks (such as I, II, IV, V, and VIII) were not so virulent. The rabbits inoculated under the skin died from 40 hours to 7 days after inoculation. In many such cases the acute septicæmia was no longer present, but a peritonitis quite invariably appeared. In the more rapidly fatal cases this was hemorrhagic; in the others a considerable amount of grayish, sometimes quite viscid, exudation had taken place. This covered the cæcum, liver, and spleen with a thin pseudo-membrane, and was found to some extent between the coils of the large intestine. The exudate consisted of leucocytes, fibrin, and immense numbers of bacteria. The relative proportions of these elements varied according to the age of the disease, the leucocytes being most abundant in the advanced stages. The swine-plague bacteria were localized more or less in the peritoneal cavity. The spleen and liver and the blood contained very few.

---

* For experiments with cultures of the German swine plague see pp. 127-129.

The course of the disease did not always correspond to this description. In a small number of instances the subcutaneous inoculation was followed not by peritonitis but by pleuritis and pericarditis. The exudate into these serous sacs was of the same nature as the peritoneal exudate in other cases. More rarely the pleuritis was accompanied by a genuine pneumonia of a croupous character.

A third form of the inoculation disease is produced by very attenuated swine-plague bacteria. This attenuation may have been due to long cultivation in the laboratory or else it may have taken place in nature. The rabbits inoculated subcutaneously may live several weeks. The place of inoculation becomes the starting point of a partly hemorrhagic, partly purulent infiltration and thickening of the skin and subcutaneous connective tissue which may extend over the entire abdomen and thorax. The skin in places may become necrotic, dry, and hard. If the animal survives, which is not infrequently the case, the inflammation becomes circumscribed into an abscess which finally heals. In these cases internal changes such as peritonitis and pleuritis are absent. In some cases, however, a parenchymatous and fatty degeneration of the heart muscle and the kidneys is present.

Such attenuated bacteria are still capable of producing all the various lesions ascribed to the more virulent varieties when injected directly into the circulation. This is very easily done by choosing a vein of the ear. After such injection we may produce a rapidly fatal septicæmia, or inflammation of the serous membranes, according to the quantity of virus injected and the relative virulence. The localization is the same as with subcutaneous inoculations. In rare cases even intravenous injections of small doses do not prove fatal, and then we may observe further localization of the virus in the joints of the extremities, chiefly the lower, or even in the subcutaneous and intermuscular connective tissue in different parts of the body. Such localizations were in a state of suppuration when they came to our notice.

It might be maintained that the attenuated swine-plague bacteria, such as those from outbreaks VIII and X, were specifically different from the very virulent varieties obtained from outbreaks VII and IX, owing to the differences in the rabbit disease produced by them. We have, however, carried out experiments which dispose of this possible objection. By making rabbits more insusceptible to the virus of swine plague by vaccination we have been able to vary the disease produced by virulent varieties so as to produce nearly all the forms caused by attenuated varieties. For example, those bacteria (IX) which produce a rapidly fatal septicæmia have, in vaccinated rabbits, produced a prolonged disease characterized in some cases only by local reaction at the point of inoculation and recovery, in others by peritonitis or pleuritis and pneumonia. The two following cases are interesting illustrations:

Rabbit No. 19 received into an ear vein about 13 cubic centimetres of bouillon culture of swine-plague bacteria, sterilized at 58° C., in five doses at periods 3 to 4 days apart. On May 26, 1891, 5 days after the last injection, it was inoculated subcutaneously with one five-hundredth cubic centimetre of a bouillon culture of living swine-plague bacteria. While a check rabbit (not treated) died within 20 hours after inoculation, this rabbit died at noon June 2, *i. e., 7 days after inoculation*. The skin and subcutis at the point of inoculation extensively thickened by purulent infiltration. *Double exudative pleuritis and pericarditis with large pneumonic regions in both lungs* in which the hepatization had advanced to the gray stage. Bits of this lung tissue examined fresh were made up chiefly of leucocytes. Swine-plague bacteria in exudate and spleen.

Two other rabbits treated with one additional injection of sterilized bouillon cultures had considerable local swelling after the test inoculation, but both recovered.

Another set of four rabbits were treated with sterilized agar cultures suspended in bouillon.* Only one succumbed to the test inoculation in 6 days, while a fresh rabbit succumbed to the same inoculation (one five-hundredth cubic centimetre bouillon culture of swine-plague bacteria from outbreak IX) in 20 hours. The lesions in the vaccinated rabbit were extensive local subcutaneous infiltration, double purulent pleuritis and pericarditis, and congestion of both lungs.

The numerous bacteria in the exudate were not attenuated, for a rabbit inoculated with a particle from the pleural cavity died within 20 hours.

An exceedingly interesting element in these results is the tendency of the swine-plague bacteria *towards the production of disease of the lungs and serous membranes* in these treated rabbits. The vaccination they have undergone has placed them nearer the pig as regards insusceptibility.

A great variety of pathological conditions is thus produced by different degrees of virulence of the same species of bacteria, the significance of which will be appreciated when we come to the disease in swine. If we would arrange the various lesions according to their fatality, the septicæmia or multiplication of bacteria in the blood stands first. Next come peritonitis, pleuritis, and pericarditis with or without pneumonia, and lastly joint disease.

The relative virulence of swine-plague bacteria from different sources thus demonstrated on rabbits has its counterpart in the relative virulence of the same bacteria with reference to different species of animals. The susceptibility of the animals used is greatest in rabbits and gray mice, and diminishes in guinea-pigs, pigeons, and fowls in the order given.

In guinea-pigs the lesions produced by subcutaneous inoculation present the various forms which we have seen characteristic of rabbits. There is this difference, however, that the more attenuated varieties are likely to have no effect on guinea-pigs beyond a slight local reaction.

---

* See p. 145 for details.

In pigeons and fowls the inoculation disease is a septicaemia, produced, however, only by the most virulent varieties of swine-plague bacteria. Attenuated varieties have no effect.

We have seen that the bacteria from outbreaks VII and IX, and from Germany, all produce a rapidly fatal septicaemia in rabbits, and in so far no difference in their pathogenic power is manifest. But the guinea-pig inoculations make a discrimination in showing that the bacteria from VII are less virulent than those from IX. Inoculation with the latter, even in very minute doses (see p. 76), produces a rapidly fatal septicaemia, and hence the effect on rabbits and guinea-pigs is alike. Inoculation with cultures from VII are on the other hand fatal to guinea-pigs only after some days, and the localizations are uncertain just as with attenuated swine-plague bacteria in rabbits. The inoculation of pigeons also shows this difference. Bacteria from VII are fatal only when the injections are made slightly into the muscular tissue. The bacteria from IX are fatal in 24 hours, even when a very minute quantity of the culture is placed under the skin.

In case of fowls neither variety of bacteria is fatal when introduced under the skin. When, however, the needle punctures the pectoral muscle the bacteria IX prove fatal in 36 to 48 hours. With pigeons the case is similar. Bacteria which fail to produce any effect when placed under the skin may still prove fatal when injected into the muscular tissue.

A further discrimination is possible in still more refractory animals, such as swine. It has been shown in the preceding pages that swine-plague bacteria from outbreak IX are the most virulent which have yet been encountered. One of the cultures of the German swine plague is still more virulent, since it proves fatal to swine after subcutaneous inoculation, while the American variety usually fails.

The nature of the pathogenic activity of swine-plague bacteria is to a certain extent cleared up by these experimental inoculations. When they are capable of multiplying in the blood they produce speedy death, probably by the development of some poison and by a modification of the blood. When multiplication in the blood does not take place, the large serous cavities still permit their growth on the lining membranes. The irritation thus set up induces a fibrinous and cellular exudation, which later on becomes completely cellular. The means by which the bacteria are destroyed in these cavities when recovery ensues is not known. Phagocytosis probably plays an important rôle here, for I have frequently seen large numbers of bacteria imbedded in the protoplasm of the leucocytes. The pleural cavity is less frequently involved. This may, however, be due in part to the fact that in most cases the inoculations were made in the region of the abdomen or the thigh.

The relative virulence of the varieties investigated thus varies considerably. Of those studied in detail recently the highest degree of virulence would belong to one form of the German swine plague. The others may be arranged in the following order : IX, VII, VIII and X.

97

THE DISEASE IN SWINE AS PRODUCED BY THE INOCULATION OF CULTURES.

In order to determine the effect of pure cultures swine were inoculated under the skin, into the veins, into the lungs, and into the abdominal cavity. They were also fed with cultures and with the viscera of small animals which had succumbed to inoculation. The details of these experiments have been given under the different outbreaks, and it now simply remains to bring the facts together.

It is at once apparent that any method of inoculation whatever can merely approximate the conditions prevailing in outbreaks of disease. The body receives in the one case a large number of bacteria at one time, while in the natural disease animals are being constantly infected by small quantities. A single small dose frequently repeated—and this is what actually takes place in an infected herd—may produce far more serious results than a single large dose. Again, the repetition of small doses may produce a disease quite different in character from that produced by a single large dose of the virus; finally, the ways in which bacteria enter the body, in the natural disease, vary greatly, and differ in most respects from the ways in which they are introduced experimentally.

These and many more considerations which need not be discussed here, serve to show that with inoculations into swine we may attempt but imperfectly two objects, first, to demonstrate that the bacteria under investigation do actually produce disease, and, second, that they may have a predilection for certain organs and tissues of the body where certain kinds of lesious are produced.

The inoculations in swine produce practically the same lesions as those which follow the inoculations of the smaller animals. Here we observe variations in the localization of the injected bacteria similar to those produced by attenuated cultures in rabbits and guinea-pigs.

Feeding cultures and viscera of inoculated rabbits has thus far proved negative.

Subcutaneous inoculation has in most cases proved negative excepting with bacteria from outbreaks I, II, and IX. One variety of the German swine-plague bacteria proved in most cases fatal after subcutaneous inoculation.* This, as has been stated before, gives them a higher potency than any American variety thus far encountered.

Injection of cultures into the circulation is usually fatal when virulent varieties are employed. These lesions depend upon the time elapsing between inoculation and death, which in turn varies according to the virulence and the quantity of culture liquid injected.†

When the animals die within one or two days the visible changes are confined to the blood, and in some cases extend to the peritoneum and

* See p. 93.
† A full account of these inoculations is given on pages 71-75.

1614——7

pleura, which are inflamed and covered with more or less exudation. If the animals live longer these changes become more accentuated, and in several cases portions of the lungs become hepatized. This hepatization may be looked upon as secondary to the pleuritis, as in case of the rabbits having similar lesions. In two cases the bacteria became localized in various joints of the body, where necrosis of bones and suppuration took place. Intermuscular collections of pus were also observed. The duration of the disease in these cases was several weeks or longer. Other organs, excepting the pericardium, in one of these animals, were not directly affected.

When the bacteria are injected directly into the lung tissue through the chest wall, death may follow within 16 to 24 hours, or life may be prolonged and severe pleuritis and pericarditis associated with hepatization of the lung tissue, may appear. Thus in pig 275 (see Plate VI) though the injection was made into the right lung, as a result the major portion of the left lung also was involved in pneumonia.

In a few experimental cases the intestines were involved. In one case, as a result of the peritonitis following an intra-abdominal injection, the walls of the small intestine were swollen, inflamed, and a copious, friable, yellowish exudate had formed on the deeply inflamed mucosa. In another case all Peyer's patches in the small intestine were swollen, very hyperæmic, and in part hemorrhagic. Intense hyperæmia of the mucosa of the stomach has been observed in several cases, both after intravenous and intrathoracic inoculation.

SWINE PLAGUE AS OBSERVED IN EPIZOÖTICS.

The variety of lesions produced by the inoculation of swine-plague bacteria is by no means so great as that observed in nature. While there are outbreaks in which considerable uniformity is observed, there are others in which each animal is a surprise to the pathologist. In general it may be stated that the lungs and the digestive tract are the chief seats of the disease, though other organs, notably the lymphatic glands, are secondarily involved. The disease is localized in the lungs and the digestive tract most likely because the bacteria gain entrance through the respiratory and digestive passages.

The lungs have been found diseased in nearly every outbreak which has been investigated. In some (notably IV, VII, and VIII) the lung disease is the predominating affection and the direct cause of death. In IX pneumonia was absent in some cases, but pleuritis and interlobular œdema were generally present. The localization of the disease in certain lobes of the lungs is quite constant. The ventral lobes are first attacked, next come the cephalic and azygos, and lastly the principal lobes. This movement of the disease seems to depend on gravity, inasmuch as the diseased is marked off from the healthy portion by a nearly horizontal line. In other words, the most dependent portions of the lungs are the ones affected first, and as the disease progresses upwards only a small portion of the principal lobe, directly under the back of the animal, remains pervious, provided the life of the animal is maintained up to this point. This localization of the disease is indicated on Plates I and II, in connection with which a description of these drawings will be found. A similar distribution of disease I have observed in cases of bovine pneumonia due to bacteria and to actinomyces, and in isolated cases of pneumonia in swine, the cause of which I was unable to trace at the time. In fact, in nearly all cases of pneumonia in swine, excepting a few which were either caused by lung worms or due to embolism, the disease involved the regions described and indicated on the plates by shading. In the exceptions the pneumonia involved portions of the principal lobes not contiguous to the ventrals.

Pathologists have defined two kinds of pneumonia, croupous and catarrhal pneumonia, or broncho-pneumonia. In the former the vesicular portion of the lung substance is chiefly affected; in broncho-pneumonia the smaller bronchioles are said to be primarily affected and the vesicular portion or alveoli secondarily. In croupous pneumonia, there is, following the stage of congestion, an emigration of red blood corpuscles, some leucocytes, and an exudate of fibrin into the alveoli. These elements are firmly matted together by the coagulating fibrin, making the diseased lung firm to the touch. In broncho-pneumonia the catarrhal condition of the smaller air tubes makes them impervious to air. The lung tissue which they supply is gradually emptied of air and assumes the appearance of red flesh, owing to the collapse of the walls of the alveoli and the distended condition of the capillary network. Subsequently the inflammation extends into the alveoli, which then become distended with cellular masses.

The definitions of pathology do not always apply to classification of lesions from the standpoint of etiology or causation. A definition from an anatomical standpoint refers to a condition. The same bacteria may under different circumstances produce a variety of conditions. Another difficulty meets us in attempting to describe the lesions due to bacteria in animal organisms. The definitions have had their origin in human diseases and are not always applicable to animal diseases. Moreover there is a difference between different species of animals. Anyone who has experimented with different animals knows that they do not react exactly alike after inoculation with the same bacteria. There seem to be certain peculiarities belonging to each species which have not yet been clearly formulated by comparative pathology. They may be due to differences in anatomical structure or to physiological peculiarities.

It is furthermore evident that the nature of the lung disease will depend more or less upon the mode of entrance of the virus. If it can enter only by way of the air tubes it will appear perhaps as a broncho-pneumonia. If it can enter the lung tissue through the circulation we may have more or less scattered centers of hepatization (embolic pneumonia). If it can enter by way of the pleura, the virus will creep along the interlobular and peribronchial tissue before it invades the parenchyma proper.

In natural infection the swine-plague bacteria enter the lung tissue chiefly by way of the air tubes. At the same time it is not improbable that occasionally they may enter the serous cavities first, i. e., invade the pleural cavities and thence the lungs. This probability is shown by our inoculations in which intravenous injections produced exudative pleuritis, and pneumonia of the most dependent portions of the lungs covered by the pleural exudate. It is not improbable that even in the natural disease the bacteria which have gained access to a portion of the lung tissue by way of the air tubes reach the pleura covering this portion, and may then by this route invade other portions of the lungs. It may be that in this way a pneumonia originally single may become double. I have also observed not infrequently that the first pneumonic infiltration of the principal lobe was at the points of contact with the diseased ventral lobe, and that the resting of a lobe against an inflamed serous surface, such as the pericardium, caused a pneumonic infiltration at the point of contact.

These facts go far to show that swine-plague bacteria may invade the lungs both from the air tubes and the pleural cavities. I am not inclined to believe that pneumonia is produced to any extent by swine-plague bacteria acting as emboli in the circulation, because they are rarely found in the blood. Comparative inoculation experiments in rabbits show that the chief indication of attenuation is inability to multiply in the blood. This probably holds with equal force in swine whose power of resistance is so much greater than that of rabbits.

We may, therefore, expect to find the character and seat of the lung lesions somewhat variable—and this is actually the case. It would be difficult to find two lungs exactly alike so far as gross appearances go. This to be sure may be due largely to the fact that animals die in different stages of the disease. Yet there are differences evidently not dependent on this fact which must be left for special pathological investigation.

After these preliminary remarks we turn to a brief description of a swine-plague lung. In general the cephalic (anterior) half is hepatized, of a dark-red or grayish-red color, firm to the touch. The pleura is more or less thickened and opaque, and covered with easily removable friable false membranes. In the more recently affected regions à faint but quite regular delicate mottling with yellow is observed to shine through the pleura when not thickened. These minute, hazy, yellowish dots usually occur in groups of four. Occasionally whitish or yellowish patches varying much in size are seen perhaps more frequently in the ventral lobes. These correspond to homogeneous dead masses of lung tissue.

When such lungs are cut into, the section presents much the same appearance, both as regards color, mottling, etc., as when viewed from the surface, excepting that the details are less distinct. In some cases in the most recently invaded territories in the principal lobe and nearer the dorsum in the other lobes, the dark or grayish-red cut surface shows grayish lines usually arranged in curves and circles. These, so far as I could determine, represent the cut outlines of the interlobular and peribronchial tissue infiltrated with cells. It has already been stated that these lines may represent the paths along which the swine-plague bacteria invade the lungs from the pleural surface.

The cut ends of the bronchi of the ventral lobes are frequently occluded with thick, whitish pus; in the other lobes a reddish froth is usually present. Rarely they also contain thick glairy mucus in which particles of dry pus and lung worms are imbedded. The contents of the air tubes in the ventral lobes may have been derived from the overdistended alveoli, or else a broncho-pneumonia, may have preceded the swine-plague pneumonia.

In microscopic sections of diseased lung tissue the alveoli and smallest air tubes are found distended with cell masses consisting chiefly of leucocytes. I have seen very little fibrin and very few red corpuscles in the alveoli, even in cases in which the disease was quite recent. It may be that the stage represented in ordinary croupous pneumonia by the presence of fibrin in connection with the cellular elements is very brief, and that it is speedily replaced by large numbers of leucocytes. The large predominance of these elements in some portions of the lungs, as well as beginning fatty degeneration, is probably the cause of the regular mottling of the lungs, as seen from the surface and above referred to. (See Plate IX, Fig. 2.) The little yellowish hazy dots represent the filled and distended alveoli surrounded by the hyperaemic walls.

The necrotic and caseous changes so frequent in swine plague deserve brief attention. The caseous changes were particularly noteworthy in outbreak IV (Plates VII and VIII, Fig. 1) and necrotic foci especially abundant in outbreak VII. (See Plate V.) The latter are usually quite small and disseminated in large numbers over the diseased lobes. The former represent larger masses from a marble to a horse-chestnut in size. There is need for a distinction between these two forms of necrosis, as I regard them as due to different processes.

The necrotic masses represent tissue which has been destroyed by the rapid multiplication of swine-plague bacteria in particular localities. Hence they are found in all stages of the pneumonia. The large caseous masses may be considered as the result of a slow death of larger territories of lung tissue, due primarily to the gradual overdistension of the tissue by leucocytes, and hence the gradual cutting off of the blood supply. One is a rapid death due directly to highly virulent bacteria, the other a slow death, or in other words a kind of dry suppuration in the later stages of the pneumonia, characteristic of the pig, and due indirectly to the irritation of perhaps more attenuated races of bacteria.

It has already been remarked that different species of animals react somewhat differently to inflammatory agents, and the tendency towards caseation in the lungs of swine is, I think, an illustration of the kind of reaction manifested by swine as a species. When irritating substances or cultures of bacteria are injected under the skin of pigs, there is a tendency to the formation of a rather firm tumor-like swelling. This tumor at first consists of a rather tough, yellowish-white mass, and only after a time softens in the center into a thick, semi-liquid pus resembling flour paste. This tendency to a firm, dry infiltration of pus is likewise observed in the sometimes quite extensive button-like necroses or "ulcers" in the large intestine. The same may be said of the large homogeneous cheesy masses into which the diseased lung tissue is converted after a time. (See Plate VII.)

It is not to be understood from the preceding that the pneumonia spreads from a single lobe, such as the ventral, in all or most cases. The case on page 69 (No. 385) is a signal illustration of the contrary. In this every lobe contained some necrotic foci which were evidently due to a deposit of swine-plague bacteria in these separate localities, and which had not yet coalesced by an extension of the disease when the animal succumbed.

The inflammation of the pleura frequently extends to the pericardium. (See Plate X.) This membrane is opaque, thickened, and its vessels distended. It may be glued to the contiguous lobes of the lungs and covered with exudate. Less frequently it is adherent to the surface of the heart, which is then covered by a false membrane, smooth or roughened, extending upon the large vessels emerging at its base.

DISEASE OF THE DIGESTIVE TRACT IN SWINE PLAGUE.

In order to elucidate this important subject we may briefly refer first to the results of inoculation, second to those cases of natural swine plague from which hog-cholera bacteria can be safely excluded.

In a considerable proportion of those animals which were inoculated with swine-plague cultures a severe catarrhal inflammation of the lining membrane of the stomach was produced. The hyperæmia was very intense, bordering on hemorrhage. Again, in another case the extension of the peritonitis, produced by intra-abdominal inoculation, along the mesentery caused a severe inflammation with exudation on the mucosa of the small intestine. In one case all Peyer's patches of the small intestine were in a hyperæmic, partly hemorrhagic condition.

In outbreak VIII there was observed in 3 out of 5 cases more or less extensive hyperæmia of the mucosa of the large intestine bordering on a hemorrhagic condition. In the case (No. 385) caused by infection from outbreak IX, the inflammatory condition of the large intestine was a prominent lesion. In these cases hog-cholera bacilli may be excluded with reasonable certainty. In the earlier cases of outbreak IV a peculiar croupous exudation appeared, which I have considered and still regard as the effect of swine-plague bacteria in the large intestine. (See Plate VIII, Fig. 2.) Of this lesion more or less has been said on page 24, in connection with a brief review of the outbreak. It will be remembered that in this epizoötic hog-cholera bacilli were found in the later cases, while swine-plague bacteria were present in a large proportion of both earlier and later cases. The croupous inflammation in this outbreak differed markedly from the necrotic and diphtheritic inflammation caused by hog-cholera bacilli. At the same time we must regard it as unsatisfactory in the present state of knowledge to reason from the pathology of swine disease to its etiology.

In the remaining outbreaks two classes appear, those in which both swine plague and hog-cholera bacilli were readily detected (II, V, VII), and those in which hog-cholera bacilli were not found, or in which their relation to the disease was highly doubtful (I, VI, IX). In most of these outbreaks the intestines were diseased, and the lesions resembled in general those found in hog-cholera epizootics. While there can be no reasonable hesitation in attributing the intestinal lesions largely to hog-cholera bacilli when they are detected, it is difficult to understand those cases in which hog-cholera bacilli are not found, or in which such attenuated forms are met with occasionally, the power of which to cause disease is highly questionable.

It is reasonable to assume, at least until more knowledge is at hand, that, even in those cases in which hog-cholera bacilli are not detected in the internal organs and yet extensive necrosis and ulceration of the large intestine is present, these lesions are due to hog-cholera bacilli or bacteria not yet recognized. The difficulty of examining the intestines for pathogenic bacteria and the amount of labor involved is very great, and hence for want of time and sufficient assistance this part of the work has been set aside in these investigations and the attention centered on the lungs and the other internal organs. It is apparent that even if certain bacteria are found in the contents of the digestive tract their relation to disease processes going on there is by no means proved, since the intestines contain a large variety of bacteria at all times. It is necessary to demonstrate that with pure cultures of such bacteria the same, or a very similar intestinal disease, can be produced. While this demonstration has frequently been made by us with virulent hog-cholera bacilli, which are also quite invariably present in the spleen, such demonstrations fail with swine-plague bacteria and with very attenuated hog-cholera bacilli, and we are at a loss to determine how much disease-producing power to attribute to them.

The production of intestinal disease by swine-plague bacteria may be supposed to go on as follows: The bacteria first attack the lung tissue and there produce more or less hepatization. The blood finds its path through the lungs partly obstructed. This reacts on the blood in the right side of the heart and the venous blood entering it. Hence there may be more or less stasis of blood in the portal circulation which in turn impairs the digestive functions of the stomach. The swine-plague bacteria in the lungs in the later stages of the pneumonia may be coughed up in the contents of the bronchial tubes, swallowed and passed through the impaired stomach unharmed into the intestines. The stagnation of feces in the large intestine furnishes the bacteria an opportunity to cause inflammation of the mucous membrane and exudation. The tendency of swine-plague bacteria to cause fibrinous inflammatory deposits on serous membranes may serve to explain such action on mucous membranes.

If we continue to follow the results of such possible localization we must assume that in the diseased intestine a considerable multiplication of swine-plague bacteria takes place, which, discharged with the feces, put into the surroundings of the swine a large number of pathogenic bacteria. These swallowed by healthy pigs may lead directly to intestinal disease without any necessarily extensive lung lesions. The virus thus multiplied by the multiplication of cases will produce a more and more virulent epizoötic in which intestinal lesions may predominate. While there is no proof that these phenomena actually take place, all the facts observed in an outbreak readily accord with such a hypothesis.

The mingling of two diseases, hog cholera and swine plague,* makes it necessary to compare briefly the pathogenic power of the bacteria causing these diseases. This mingling has greatly complicated our understanding of the extent and the kind of lesions attributable to each bacterium. Thus, in hog cholera, the chief force of the disease is spent upon the digestive tract. The lungs are frequently involved in collapse and broncho-pneumonia of limited extent, but collapse seems to be not uncommon in apparently healthy pigs, and broncho-pneumonia may be conceived of as developing from collapse without the necessary intervention of disease germs. Again, the presence of lung worms may account for much of the localized bronchitis and hepatization. It is easy to understand that hog-cholera bacilli accidentally present in the blood in disease may pass through a healthy lung without injury, while lungs affected with collapse and lesions due to lung worms may offer a favorable opportunity for the lodgment of emboli containing bacilli. The disease process thus starting up may be continued by sputum bacteria (resembling or identical with swine plague). It becomes difficult therefore to determine whether hog-cholera bacilli have any specific effect on the lungs.

In swine-plague the exact reverse is true. The predilection of swine-plague bacteria for the lungs is a demonstrated fact even with small experimental animals. Their exact effect on the mucous membranes of the caecum is not easily determinable owing to the frequent association with hog-cholera lesions. That they produce a diffuse intense inflammation, associated at times with fibrinous exudation, will not, I think, be denied in view of the facts brought out in the experimental part of this report.

---

* See also p. 138 for some practical observations on this subject.

We have thus in hog cholera primary disease of the intestines with perhaps secondary localizations in the lungs; in swine plague primary disease of the lungs with secondary infection of the intestines.

The pathological action of these two kinds of bacteria can not be formulated with any precision without extended investigations directed to that subject alone. Meanwhile a few broad facts can be deduced from the inoculation experiments. Hog-cholera bacilli multiply in clumps in the capillaries of the parenchyma of the various organs, where they primarily obstruct the circulation, and thus produce necrosis of tissue in their immediate vicinity. When this takes place in the follicular apparatus of the intestine (in rabbits) necrosis of portions of the mucous membrane, followed by ulceration, may ensue. They do not produce fibrinous and cellular exudation on serous membranes, and probably do not multiply on these membranes. Secondarily, they produce parenchymatous degeneration of the liver, kidneys, and heart-muscle, which degeneration may be due to the toxic bodies formed by them.

Swine-plague bacteria, on the other hand, seem to multiply diffusely and abundantly wherever such multiplication takes place. When in the blood death is exceedingly rapid; when in the large serous cavities, exudates are quickly produced and death may ensue in from 3 to 7 days; when under the skin necrosis and suppuration take place.

The following table gives briefly the important differences observed between hog-cholera and swine-plague bacteria. Those features common to both are omitted :

SWINE-PLAGUE BACTERIA.	HOG-CHOLERA BACILLI.
Morphological.	*Morphological.*
1. About 1 μ long and 0.5 to 0.6 μ wide in (cover-glass preparations mounted in balsam).	1. About 1.2 to 1.5 μ long and 0.6 μ wide.
2. They show the so-called polar stain in certain conditions.	2. No characteristic polar stain. The central part of rod frequently less stained than periphery.
3. Non-motile.	3. Motile (possess flagella).
Biological.	*Biological.*
4. Growth in bouillon feeble.	4. In bouillon moderate.
5. Growth on gelatine feeble or absent.	5. On gelatine more vigorous than swine plague.
6. Growth on potato fails.	6. On potato usually abundant.
7. Tends to produce acid reaction of culture liquid.	7. Tends to produce alkaline reaction.
8. Produces no fermentation of glucose.	8. Causes fermentation of glucose with production of CO_2 and H.
9. Produces phenol and indol (?).	9. Produces neither phenol nor indol.
10. Rapidly destroyed in water, in soil, by drying.*	10. Quite resistant to the same agents.*

* See page 92 for table.

Pathogenic.	Pathogenic.
11. Multiplies diffusely in blood or on serous membranes.	11. Multiplies in blood vessels in clumps.
12. Produces septicæmia; fibrinous and cellular inflammation of serous membranes and pneumonia in small animals.	12. Does not produce inflammation of serous membranes. Produces parenchymatous degeneration of vital organs. Necroses in liver. More attenuated varieties cause infiltration and ulceration of Peyer's patches and infiltration of lymph glands.
13. Most virulent varieties are fatal to mice, rabbits, guinea-pigs, and pigeons in very minute doses.* Death in 16 to 20 hours.	13. Most virulent varieties are fatal to mice, rabbits, and guinea-pigs in minute doses.* Death in 5 to 8 days.

In view of the mingling of these two diseases can we by inoculation of both bacteria at the same time gain any information as to their relative activity? If both kinds of bacteria were of unchanging virulence this might be done, but we may have in natural outbreaks attenuated swine plague associated with virulent hog-cholera bacteria and the reverse. Or one kind of bacteria may invade the animal after the other has spent its energy. The variety of combinations which may occur in nature is too great to be imitated by experiment.

It was, nevertheless, desirable to see what effect the inoculation of mixed cultures might have. Hence the following experiment was tried, the result of which, though very important, is not conclusive.

May 12, 1891. Cultures of swine plague from outbreak IX and of hog cholera from a Virginia epizoötic, both the most virulent in the laboratory, were chosen. From the growth on an agar culture of these swine-plague bacteria 24 hours old a turbid suspension in bouillon was prepared. A bouillon culture of the hog-cholera bacilli only 24 hours old in which the growth was abundant was also on hand. With these cultures three pigs, about 6½ months old, of the same breed and lot, were inoculated as follows:

No. 462 received into the right lung one-fourth cubic centimetre of the swine-plague suspension, and three-fourths cubic centimetre of the hog-cholera bouillon mixed.

No. 463, inoculated in the same way with twice this quantity, i. e., one-half cubic centimetre swine-plague and 1½ cubic centimetre hog-cholera culture.

No. 461 received 5 cubic centimetres of the hog-cholera culture alone into the right lung.

No. 463 was dead next morning, within 16 hours after inoculation. Animal in good condition; weight 72 pounds. Subcutaneous fat reddened. Blood oozes from the cut vessels and coagulates feebly. Arborescent injection of minute bloodvessels of connective tissue in axilla and between muscles on thorax. In abdominal cavity a considerable quantity of yellowish serum. Ventral portion of spleen imbedded in an elastic whitish exudate. Costal and diaphragmatic pleura show intense vascular injection. In right pleural sac about 140 cubic centimetres blood-stained

* It must be borne in mind that the various attenuations of hog-cholera and swine-plague bacteria can not be individually considered with reference to their pathogenic effect in a tabulated form; we must refer to the text for these.

liquid, in left 70 cubic centimetres. Lungs but partly collapsed : œdematous. Over the entire right lung a very thin friable exudate; over the left this is found only on most dependent portion. Right lung punctured by needle in principal lobe. The puncture shows as a hemorrhagic spot.

In the stomach—which is filled with food—and in intestines, nothing abnormal. Kidneys with base of pyramids dark red. Spleen slightly enlarged.

In the spleen pulp only swine-plague bacteria detected. In an agar culture therefrom no motile bacilli seen. A bouillon culture from the peritoneal exudate contains only swine-plague bacteria. Cover-glass preparations from both pleural and peritoneal exudate show numerous polar stained bacteria.

No. 462 manifested labored breathing soon after the inoculation. It lay on its bedding unable to get up. It was found dead May 16.

Considerable reddening of the skin on ventral aspect of body. Subcutis as in 463. In abdominal cavity a small quantity of yellowish serum and some fibrinous coagula. Spleen quite large, congested and friable.

Thoracic cavity. Right pleural sac contains 100 cubic centimetres of turbid dark red fluid. Considerable friable exudate covering costal and pulmonary pleura of this side, especially abundant on small lobes. Pleura opaque, thickened and wrinkled. Both cephalic and ventral lobes not enlarged, but firm, on section grayish red, solid. Bronchioles filled with whitish muco-pus. The needle puncture in principal lobe is the center of a hepatized mass, fully 2 inches in diameter; on section some lobules bright red, others hemorrhagic, dark red. Nearer periphery of this mass the hepatization is grayish red ; the interlobular and peribronchial tissue appear as irregular grayish lines. In the left lung the principal lobe congested; slight interlobular œdema. Along cephalic border, under pleura, a dark red hepatized mass, about one-half inch in diameter. Tip of ventral lobe adherent to pericardium, beneath adhesions small foci of dark red hepatization. In terminal bronchi adult lung worms.

Pericardium adheres to heart surface by means of a thick yellowish pseudo-membrane. Glands along posterior aorta hemorrhagic.

Digestive tract : Stomach empty. Mucosa thrown into folds and covered with an abundance of very viscid mucus. In fundus it is congested. Hyperæmia also around cardiac expansion. Duodenum pigmented. Mucosa of ileum swollen and sprinkled with small hemorrhagic spots. Mucosa of cæcum and upper colon discolored. In lower colon minute whitish spots, with injected margin, evidently superficial ulcers. Meso-colic glands enlarged, hyperæmic.

Liver somewhat enlarged, quite firm. Acini slightly projecting; some dark red; in most of them intralobular necrotic masses, the result of some former disease.* Portal glands enlarged and hyperæmic. Kidneys congested; pyramids dark bluish red. Glomeruli just visible to naked eye as injected points.

Bacteriological examination : In cover-glass preparations from the hepatized mass of right principal lobe very many swine-plague bacteria and a few somewhat larger forms. In two agar plates from the same region large number of colonies of two kinds appear. One represents swine-plague about ten times more numerous than the hog-cholera colonies, which are on second day twice as large. In cover-glass preparations of spleen pulp a considerable number of bacteria appearing as hog-cholera bacilli. The same is true of the liver. Agar plates from each organ contain only colonies of hog-cholera bacilli.

No. 461, though sick for a few days, fully recovered.

This experiment illustrates the greater rapidity of action of swine-plague bacteria even in small doses. It also demonstrates their greater virulence, for the pig inoculated with a large dose of hog-cholera bacilli

---

* This condition of the liver will be discussed in a subsequent publication. It bears no relation to the inoculation, since it is found in swine otherwise healthy.

alone, although sick for a few days, recovered. Another fact of importance illustrated in these cases is the slow action of hog-cholera bacilli in invading other organs. Thus the swine-plague bacteria had killed No. 463 before the hog-cholera bacilli had invaded the spleen and liver. In No. 462, however, enough time had elapsed to permit the hog-cholera bacilli to spread through the body while the swine-plague bacteria were now limited to the lungs. The invasive power of hog-cholera bacilli, though slower, is nevertheless more lasting.

There is no evidence to support the view that either swine-plague or hog-cholera bacilli may produce serious disease in man. No bacteria have thus far been described as the cause of human disease which are identical with either of the bacteria of swine disease. Nevertheless the possibility of an occasional transmission from animal to man can not be wholly set aside until more thorough bacteriological investigations of human diseases shall have been made in those localities where infectious swine diseases are very prevalent.

It is of interest to note that among human diseases typhoid fever bears a close resemblance to hog cholera, not only as regards the general character of the specific bacteria, but also with reference to the disease itself. Again the diplococcus of croupous pneumonia in man has many points of likeness with swine-plague bacteria. The general pathological effect as well as the tendency to produce various forms of disease by localizations, such as pleuritis, pericarditis, meningitis, are strikingly similar to the miscellaneous lesions caused by swine plague bacteria. Typhoid fever and pneumonia are not infrequently associated in man, thus affording another point of similarity to the frequently associated swine diseases. Another peculiarity which is common to the pneumonia bacteria in man and to those of swine-plague is their frequent occurrence in the mouth and upper air-passages of man on the one hand, of swine on the other.*

---

* See the following chapter and appendix.

ATTENUATED SWINE-PLAGUE BACTERIA IN SPORADIC CASES OF PNEUMONIA AND IN SEPTIC DISEASES OF SWINE, AND IN THE UPPER AIR PASSAGES OF HEALTHY SWINE AND OTHER DOMESTICATED ANIMALS.*

During the investigations of the epizoötic forms of swine plague described in these pages, bacteria not distinguishable from those of this disease were occasionally encountered in sporadic cases of pneumonia. The affected swine were usually those which had been exposed to hog cholera or swine plague, or mixed outbreaks, or had been inoculated and had survived. They were kept isolated for a time by themselves or with other cases of similar history, and they generally died after some months or became so unthrifty that they were killed. Occasionally they died with some inflammation of the serous membranes, such as peritonitis, pericarditis, or pleuritis. In such cases when examined swine plague bacteria were as a rule detected. These lesions were attributed to injuries which the swine, penned together, inflicted upon one another by fighting. If we bear in mind the variety of lesions which may be produced by swine plague bacteria when inoculated into healthy swine (pneumonia, pleuritis, pericarditis, peritonitis, inflammation, and suppuration of the joints, inflammation with exudation in the intestines), we need not be surprised to encounter them in similar affections arising under the usual conditions.

The sporadic cases of pneumonia were puzzling in so far as it was difficult to account for the presence of swine-plague bacteria. These, as we have shown, very easily perish by drying, and when in water and in media unsuitable for their development. How they can survive for weeks and months in the surroundings of swine is inexplicable. To find some other explanation I was led to examine the respiratory tract of apparently healthy pigs to see if here, in a manner analogous to the bacteria of pneumonia in man, they survived and became a potential source of infection under certain conditions. The result was positive in many cases. Swine-plague bacteria or bacteria not distinguishable from them were found in the nose, at the base of the tongue and in the larynx of a certain number of healthy swine.

---

* In this chapter no effort has been made to give an historical review of those experiments made by former observers which demonstrated the occasional virulence of the saliva of man and animals. The general relationship of these bacteria from various domesticated animals to swine plague bacteria is, I think, brought out for the first time in these pages. See also appendix to this report.

The cases which have come under our observation may be grouped into several classes :

1. Sporadic pneumonia.
2. Other affections, chiefly inflammations of serous membranes.
3. Older animals having passed through diseases but apparently healthy.
4. Younger animals not known to have been exposed to disease.
5. Animals entirely free from these bacteria.

Groups 1, 2, and 3 generally include, in our observations, older animals.

The method used in all cases was the subcutaneous inoculation of rabbits with the tissue, secretion, or exudate suspected of carrying these bacteria. The inoculations were made in part conjointly with Veterinarian F. L. Kilborne; in part by him alone. Great care was taken to prevent infection from external sources. At first only the nasal secretion was used. This was drawn up with a sterile pipette and then introduced into the subcutis of the rabbit through a small incision in the skin. This method did not seem quite satisfactory, and in a number of cases the animal was killed by bleeding, the mouth carefully exposed from below and mucus collected from the base of the tongue, the posterior nares, epiglottis, and from beneath the vocal cords. The mucus was placed in a sterile watch glass. The skin of the rabbit was freed from hair and thoroughly washed with a solution of mercuric chloride 1:500; with flamed scissors, the skin was cut through usually on the side of the abdomen, a little pocket formed, and a drop of mucus put into it with flamed forceps. The incision was either left to itself or united with a stitch of sterilized silk.

These precautions thus prevented any external infection. At the same time a prolonged experience in such work leads me to believe that swine-plague bacteria are not transmitted in a dried condition, excepting, perhaps, for a few days only. The outcome of the inoculations also inclined us to exclude any accidental infection. When several rabbits were inoculated with mucus from the same locality they either died together or both survived.

The demonstration of swine-plague bacteria in the upper air passages of swine naturally led to an examination of the air passages of other domesticated animals by the same methods. These are reported by Dr. V. A. Moore, assistant in the laboratory, to whom this part of the work was intrusted. It will be seen from this report that an attenuated variety of bacteria, belonging to the group of swine-plague bacteria and not distinguishable from them, inhabit the mouth and upper air passages of such domesticated animals as cattle, dogs, and cats, and that some sporadic diseases among smaller animals, such as rabbits, guinea-pigs, and fowls, are caused by the same bacteria. This group has a wide distribution, therefore, and may be regarded as a more or less common inhabitant of the mucous membranes.

In the following brief synopsis of the cases illustrative of the above statements, they are grouped as nearly as possible in accordance with the classification above outlined:

1. *Sporadic pneumonia.*—Pig No. 481. Placed in infected hog-cholera pen January 11, 1888. Since February 1, large sores are observed on the shoulders, attributed to bites inflicted by other swine in the pen. It died March 5, very much emaciated. In the cæcum several crater-like ulcerations, involving the muscular coats. In the upper colon about a dozen ulcers, one-fourth to three-eighths inch in diameter.

The ventral and cephalic lobes of both lungs and the roots of principal lobes, also foci in median lobe, hepatized, pale red.

A bit of diseased lung tissue torn up in sterile bouillon and one-fourth cubic centimetre injected into a rabbit. It died in 40 hours, with exudative peritonitis, due to swine-plague bacteria.

In this case these bacteria may have come from the other animals in the infected pen. In the following case the source is not so obvious.

No. 492, received at the Station January 25, 1888, with three others (490–493, inclusive). In making inoculations from the nasal mucus of these pigs, to be described later, Dr. Kilborne noticed that the secretion was muco-purulent, while in other pigs handled at the same time it was serous. No. 492 became unthrifty, and died about a month after it was brought to the Station.

The lungs were firmly bound to diaphragm and chest wall by old adhesions. The ventral and cephalic lobes, the roots of the principal lobes of both lungs and the median lobe airless, in condition of broncho-pneumonia. The catarrhal masses in the alveoli appeared as a pale grayish yellow, delicate mottling under the pleura. Throughout the hepatized portions are yellowish necrotic foci one-sixteenth to one-eighth inch in diameter. The smaller bronchi contain a soft, creamy muco-pus. In the large bronchi, the mucosa hyperæmic. Bronchial glands enlarged, very pale and firm.

Two rabbits inoculated as in preceding case. One died in 40 hours, the other on the third day. Bacteria showing the polar stain present in large numbers in organs of the first; in small numbers in the second rabbit. In cultures they are not distinguishable from swine-plague bacteria.

Pig No. 267, received October 16, 1889. This animal belonged to a lot of swine which had been inoculated with attenuated hog-cholera cultures and subsequently exposed to hog cholera. The animal died January 2, 1890. In the lungs were regions of collapse and broncho-pneumonia, as well as marked bronchitis. From the abundant mucus in the upper trachea two rabbits were inoculated. They succumbed on the fourth and the sixth days, respectively, to an exudative and hemorrhagic peritonitis characterized by the presence of swine-plague bacteria.

Pig No. 188. This animal was placed in an infected hog-cholera pen March 13, 1889, and removed in 5 days. It was killed December 9, after a long period of unthriftiness. The ventral lobes of both lungs were hepatized, of a grayish-red appearance (broncho-pneumonia). The liver covered with bluish and whitish patches, in which the interlobular tissue is thickened. This thickening is limited to the surface. One rabbit inoculated with mucus from larynx died in 10 days, with extensive subcutaneous purulent infiltration and exudative peritonitis. The only bacteria found were the swine-plague bacteria.

Pig No. 308 was purchased from a neighboring farm February 7. Four days later it was killed and hepatization of a portion of both principal lobes determined, together with bronchitis of the air tubes leading to these diseased regions. Two rabbits were inoculated from the tracheal and pharyngeal mucus, and one from the hepatized lung. The latter remained alive, while the two former died in 3 and 4 days, respectively, with extensive exudative peritonitis. The swine plague bacteria alone were present in the organs of both rabbits.

Pig No. 119 was placed in an infected hog-cholera pen January 22, 1889, where the swine-plague disease was present. September 2 it was transferred to another pen, and killed January 13, 1890. The internal organs were healthy, with the exception of the lungs, in which there were several small foci of collapse and some lung worms. One rabbit inoculated with nasal mucus from the base of the tongue died in 3 days. There was considerable phlegmonous inflammation and peritonitis as the result of the inoculation. The peritoneal exudate contained immense numbers of swine-plague germs. The other rabbit inoculated with mucus from below the vocal chords died in 3 days with the same lesions and bacteria present.

2. *Other affections.*—Pig No. 180. This pig was exposed to hog cholera in an infected pen March 13, removed March 18. It survived the exposure and was killed June 4. Its internal organs were normal with exception of the peritoneal cavity, which contained a considerable quantity of cloudy serum, the result of peritonitis. Four rabbits were inoculated from mucus obtained from below the vocal chords and at the base of the tongue. The two inoculated from the latter source died in about 24 hours. In both the spleen and blood contained large numbers of swine-plague bacteria. The two inoculated with tracheal mucus died within 36 hours. The spleen of both contained large numbers of the swine-plague bacteria.

Pig No. 202. This animal was fed in May, 1889, with small quantities of culture liquid containing hog-cholera bacilli twice a day for a week without producing a fatal disease. November 15 it was penned with No. 119, which harbored the swine-plague germs. February 14 it was killed by bleeding, and two rabbits inoculated with mucus from the base of the tongue and from the larynx respectively. Both rabbits died, one in 3, the other in 4 days. The lesions consisted in subcutaneous inflammation and exudative, partly hemorrhagic peritonitis. Swine-plague germs were present in the various organs, especially in the peritoneal exudate.

No. 37. This animal had been inoculated in the lungs with a culture of swine-plague bacteria, October 8, 1888. January 27, 1889, it was exposed to hog cholera, which exposure it survived. It was killed August 16, in a very good condition. The only noticeable abnormal condition was the firm adhesion of the right lung to the chest wall by means of firm, fibrous tissue, the result of the swine-plague inoculation. The lungs themselves healthy. Three rabbits were inoculated with mucus from the respiratory passages (tongue, nose, and larynx). All three rabbits died, two in 48 hours, the third in 4 days. In the latter the inflammation at the place of inoculation and the peritonitis were most pronounced. In the organs of all three rabbits the swine-plague bacteria were present, especially numerous in the peritoneal exudate.

The following case is of importance, because the bacteria obtained from it were tested upon pigs and found virulent.

No. 164 survived a hog-cholera exposure early in 1886, and in June of the same year was transferred to another pen. It was kept until October, 1887, at which time it died of peritonitis, probably occasioned by injuries received in fighting with other pigs in the same pen. No bacteriological examination was made of the peritoneal exudate, but two rabbits were inoculated with mucus from the trachea.

One died in 20 hours; in the various organs numerous bacteria showing polar stain. The other died on second day, with the same bacteria, as demonstrated by the microscope and cultures in bouillon and gelatine.

To test farther the relative virulence of these bacteria a peptone bouillon culture was prepared, of which two mice received subcutaneously one-twelfth cubic centimetre; two pigeons received subcutaneously and into the muscular tissue one-fourth cubic centimetre; one guinea-pig received subcutaneously one-eighth cubic centimetre, and one rabbit one-twelfth cubic centimetre.

The guinea-pig and the pigeons remained well. One mouse died in 24 hours, the other in 3 days. In the first large numbers of bacteria; in the second few. The rabbit died within 2 days. In this animal there was slight infiltration of skin and subcutis at point of inoculation. Peritonitis. Invagination of lower colon. In the various organs and peritoneal exudate numerous bacteria showing polar stain.

The pathogenic character of these bacteria was further demonstrated by inoculation of swine.

November 11, 1887. No. 431, 6 weeks old, received into right lung through chest wall 2½ cubic centimetres of a peptone bouillon culture.

No. 432, 6 weeks old, received the same.

No. 433, 6 weeks old, received 5 cubic centimetres subcutaneously.

No. 433, though affected by the inoculation for a time, recovered.

No. 431 appeared paralyzed and unable to get up 3 or 4 days after the injection. Respirations somewhat quickened. It took very little food. Found dead 14 days after inoculation.

Slight infiltration in subcutis at point of injection. Right lung collapsed. Pleural cavity half full of blood-stained serum. Considerable spongy, yellowish exudate is loosely attached to the walls of the thorax, the lung surface, and the diaphragm. The lung tissue is not hepatized excepting a small mass which is necrosed and which probably represents the place where the needle penetrated. Left lung not affected, closely bound to thorax by fibrous adhesions which give way without much difficulty. The lymph gland near root of neck very large, whitish on section, small yellowish foci in cortex and medulla. Slight fibrinous exudate and considerable yellow serum in abdomen. In upper and middle portion of colon, the mucosa is covered by patches of a very thin grayish deposit, suggesting necrosis of the surface epithelium.

Cultures from spleen negative; those from pleural exudate show the injected bacteria only. A rabbit inoculated subcutaneously with about one drop of serum from the pleural cavity diluted in sterile beef infusion died within 48 hours. No local reaction or peritonitis. Innumerable polar-stained bacteria in the spleen, which is enlarged, friable. Fewer in the liver; still fewer in blood from heart. Cultures revealed the same organism.

No. 432 breathed with difficulty for several days after the injection. It seemed feverish and refused food. Within two weeks it was greatly improved. December 27 its rectum was prolapsed and it died a week later. At the autopsy the cause of death was found to be invagination and rupture of ileum. The lungs presented some interesting features. The right lung was adherent to thoracic walls and diaphragm by means of fibrous tissue not yet very firm. The left lung was adherent in several places. The various lobes of the right lung were bound together by fibrous tissue and to a tumor lying between principal and ventral lobe along ventral border of lung. The tumor was removed by careful dissection, the lung tissue being slightly condensed and hyperæmic near attachment. There was no hepatization of either lung. The tumor felt tense, walls about one-eighth inch thick, inside dark red. Contents putty-like, grayish, made up of pus. Pericardium thickened and attached in several places to epicardium, which is likewise thickened, opaque.

In a gelatine-roll culture inoculated with a particle of the dry pus numerous colonies of the injected bacteria appeared after a week. A rabbit inoculated with a particle of pus died in 5 days with considerable local infiltration and enlarged spleen. No peritonitis. In the spleen a moderate number of bacteria identified in cultures with those originally injected into the pig.

These bacteria were in every way like the various swine-plague bacteria with one exception. In liquid cultures when one or two days old, translucent capsules could be seen surrounding the bacteria individually when the liquid was examined in the hanging drop and the border of the drop was carefully scrutinized. This method I have found of much service in disclosing the presence of these glassy envelopes when drying and staining failed to bring them into view. The bacteria, as they are drawn to the border of the drop, do not touch one another, but remain separated from each other by a space of definite width. Careful focussing then will also bring out the very faint outlines of the oval transparent capsules. In the inoculations above described, the capsules served as an important means of identifying these swine-plague bacteria from case to case.

3. *Older animals apparently healthy but previously exposed to disease.*— Under this head would come some of the cases already recorded and the following :

No. 420 had been inoculated with hog-cholera bacilli October 6, 1887, and with swine-plague bacteria October 20. March 1, 1888, a drop of mucus taken from nares with a capillary pipette and forced into subcutis of a rabbit through a skin incision. The rabbit died in 10 days with exudative pleuritis, the exudate containing swine-plague bacteria.

No. 219. This animal was inoculated subcutaneously with attenuated hog-cholera cultures September 27, 1889. It was killed January 13, 1890, and found normal. There was at the place of inoculation an encysted caseous mass about 1 inch in diameter. Three rabbits were inoculated with mucus from the respiratory tract. Of these but one rabbit died on the 11th day with purulent pleuritis and pericarditis. Swine-plague bacteria were obtained from the exudate and their virulence tested by inoculating a fresh rabbit with a pure culture.

4. *Animals not known to have been exposed to disease.*

October 6. Nasal mucus obtained from the nares of a healthy pig on a flamed glass rod is stirred up in sterile water and one-half cubic centimetre injected subcutaneously into two rabbits.

No. 1. Dead October 12. Purulent thickening of the subcutis at point of inoculation and extending thence over abdomen and thorax as a sanguinoleut effusion. Peritoneum roughened. A cover-glass placed on cæcum, removed and stained, shows immense numbers of bacteria exhibiting the polar stain. The same bacteria scarce in blood, spleen and liver.

No. 2. Dead October 13. Lesions as in No. 1. Exudative peritonitis with ecchymoses on cæcum.

Cultures from both cases on gelatine and in bouillon contain only swine-plague bacteria.

Additional inoculations were made February 2 and March 1, 1888. Nasal mucus from supposedly healthy pigs was collected in a capillary pipette and a drop forced with a rubber bulb into a subcutaneous pocket made by an incision through the skin. The incision was closed with collodion.

Date.	Rabbit No.	From pig No.	Result.
Feb. 2	1	483	Negative.
Feb. 2	2	484	Do.
Feb. 2	3	493	Dies in 48 hours. Large numbers of polar-stained bacteria in blood, spleen and liver.
Feb. 2	4	491	Negative.
Mar. 1	5	490	Do.
Mar. 1	6	491	Dies in 7 days. Extensive purulent infiltration of subcutis; cultures negative.
Mar. 1	7	493	Negative.
Mar. 1	8	483	Do.
Mar. 1	9	482	Dies in 13 days. Pneumonia and exudative pleuritis.
Mar. 1	10	484	Dies in 7 days. Same as No. 6. Bacteria obtained from spleen in cultures fatal to a rabbit in 48 hours.
Mar. 1	11	468	Dies in 13 days. Same as No. 9.

No. 468 belonged to one lot brought to Station January 4, 1888.

Nos. 490 to 493, inclusive, to another lot brought to Station January 25, 1888.

Nos. 481 to 484, inclusive, to still another lot brought to Station January 10, 1888.

The inoculation disease in the rabbits varied considerably. In one, death ensued in 48 hours. Nos. 6 and 10 died in 7 days, and in both there was extensive purulent infiltration of subcutis over abdomen and thorax. Finally, in Nos. 9 and 11 there was exudative pleuritis with pneumonia. In No. 9, one-half, in No. 11, the whole of the left lung hepatized. In the pleuritic exudate and the lung tissue large numbers of polar-stained bacteria.

None of the three lots of pigs were thus free from these bacteria. It will be noticed that the mucus from four pigs was tested twice and that the results were not uniform. No animal produced disease more than once while one failed to produce disease both times.

Pig No. 303. January 28, 1890. This animal was taken from a herd which had been purchased January 25. The farm from which it came is said to have had some form of swine disease on it nearly a year ago. The thoracic organs of the pig were normal in appearance. Two rabbits were inoculated, one with mucus from the larynx, the other with mucus from the pharynx. Both succumbed on the third and fourth day, respectively, to exudative peritonitis. In the exudate the swine-plague bacteria were quite abundant.

December 21, 1890. No. 447, about 4 months old, received yesterday with a lot of other pigs from a farm which has been free from disease for several years. The pig was killed by bleeding, and although it seemed well, the anterior half of both lungs was found diseased. The diseased lobes were of a pale red appearance and seemed œdematous rather than hepatized. There was considerable muco-purulent secretion in the smaller bronchi. In the terminal bronchi of the principal lobe some lung worms.

One rabbit was inoculated subcutaneously with a particle of lung tissue and two with mucus from base of tongue. The former remained well; both the latter died. One died within 40 hours with considerable local purulent infiltration of skin and subcutis. Spleen enlarged, containing bacteria showing polar stain. Culture on agar from spleen contains only swine-plague bacteria. The second rabbit died in 3 days, with very extensive subcutaneous infiltration and exudative peritonitis. Spleen enlarged and dark colored. In the organs few bacteria, in the exudate a large number. Cultures from spleen and exudate contain only swine-plague bacteria. In order to test the virulence of these bacteria in pure cultures, an adult rabbit was inoculated subcutaneously with an equivalent of one five-hundredths cubic centimetre bouillon culture diluted in bouillon. The rabbit died on the fourth day with lesions like those of the preceding case.

5. The results thus far obtained must not lead us to infer that all swine carry with them bacteria closely allied to swine-plague bacteria. Some herds are entirely free from them, as the following statements will show :

Pigs Nos. 116 to 133 were received January 4, 1889, from a place in the District of Columbia where swine diseases have not prevailed for a number of years, and where much care is bestowed on the rearing of swine. From four of these, nasal mucus was collected and four rabbits inoculated subcutaneously with one or more drops. All remained well.

From another lot of eight, received February 4, two were tested. The results were likewise negative.

To test the matter somewhat more thoroughly in subsequent cases, the animals to be examined were killed and mucus taken from various places in the upper air passages.

No. 205, one of a lot regarded healthy, killed June 15, 1889. Pericardium firmly adherent to the heart. Three rabbits inoculated with mucus from posterior nares, the base of tongue and from below vocal cords. All remained well.

No. 207, from another lot, killed on the same day. The organs were in general healthy with the exception of the large intestine. The mucous membrane appeared to be in a state of mucous degeneration. It was swollen, partly translucent. On valve and along colon smal' patches of a thin, friable, yellowish deposit. Five rabbits were inoculated—two from this deposit, three with mucus from trachea, posterior nares, and base of tongue. All remained well.

It would be going too far to maintain that all forms of lung disease were the result of the invasion of swine-plague bacteria. The absence of these bacteria is well illustrated by an outbreak of hog cholera investigated in the fall of 1887, and reported in the special report on hog cholera, pp. 39–52.

In about one-half of the fifty cases there was some disease of the lungs. This was in part simple collapse, in part broncho-pneumonia involving one of the small ventral lobes. Of the sixteen rabbits inoculated with particles of diseased lung tissue from sixteen cases, four survived and the remainder died of hog cholera. Swine-plague bacteria were not detected. It is reasonable to assume that if they had been present in the upper air passages they would have sooner or later invaded the diseased lung tissue and appeared in the inoculated rabbits.

In addition to the foregoing experiments a few inoculations into rabbits were made with mucus from the cæcum of healthy pigs, but they were negative so far as swine-plague bacteria are concerned. The mucus was taken from the crypts on the ileocæcal valve and the surface of the patch in which they are imbedded.

Cæcum No. 1. One rabbit and two mice inoculated subcutaneously. No result.

Cæcum No. 2. One rabbit inoculated. Died in 8 days. Extensive sanguinolent and purulent inflammation of the subcutis of abdomen. Peyer's patches swollen and pigmented. The appendix of cæcum swollen, blackish; ulcers on mucous surface. Cultures from internal organs wholly negative.

Cæcum No. 3. One rabbit inoculated. Died in 8 days. Subcutis and appendix as in preceding case. Cultures sterile.

Cæcum No. 4. One rabbit inoculated. Died in 11 days. On thigh, an abscess between muscles half as large as a hen's egg. Center disintegrated into a curdy mass. No other lesions. In the pus of abscess large numbers of bacilli of various lengths, staining feebly. They fail to grow in culture media. Internal organs free from bacteria.

It may be claimed that the presence of swine-plague bacteria on the mucous membranes of healthy swine and other domesticated animals is an argument against the specific character of swine-plague bacteria, and hence against the existence of a specific disease induced by them. We have already met this argument by the numerous successful inoculations of swine-plague bacteria into healthy swine, by which a disease like the natural disease has been produced. The attenuated virulence of the bacteria in the air passages makes it probable that few of them are able to produce disease excepting in a secondary rôle.

There are two infectious diseases in man which in this respect offer some striking analogies to swine plague. The disease known as croupous pneumonia is chiefly associated with bacteria (*Diplococcus pneumoniæ*) bearing much resemblance to swine-plague bacteria. These bacteria are now regarded as the chief, if not the only, cause of pneumonia. Strangely enough, bacteria not distinguishable from these are occasionally encountered in the saliva of healthy persons. Senator, Pasteur, and Sternberg were among the first to call attention to the fact that rabbits inoculated with sputum may die of septicæmia, and the bacteria found in the internal organs of these rabbits were identified subsequently with the bacteria of pneumonia. We have thus a complete analogy between swine-plague and croupous pneumonia.

In another disease facts of similar nature have recently come to light. Roux and Yersin* have found in the mouths of about 10 per cent. of all healthy persons examined bacilli which have no pathogenic effect, but which resemble closely the bacilli of diphtheria, which, in fact, they regard as very attenuated forms of the diphtheria bacillus. Löffler and others had previously found similar bacilli in the mouths of healthy persons.

There are several important questions raised by the discovery of attenuated forms of disease-germs in the surroundings, and in the body of animals. Do these bacteria belong to the same species as the virulent forms, and, if so, can they gain virulence enough under certain circumstances to produce disease? That pathogenic bacteria may gain in virulence has been shown by Pasteur, Roux, and others. Whether this applies to all kinds of disease-germs may be reasonably doubted, and the experiments thus far tried by us to increase the virulence of attenuated hog-cholera and swine-plague bacteria have shown that by passing them through susceptible animals no decided increase in virulence is observed. On the other hand, so far as methods are able to inform us, these same attenuated bacteria found in the air passages of healthy animals, and the swine-plague bacteria proper found in disease, do belong to the same species, and must be regarded as simple varieties.

These discoveries also point out that the one property of pathogenic bacteria which must receive special attention is their *relative virulence*. This seems to be the one factor which determines the severity and the communicability of infectious diseases.

---

* Annales de l' Institut Pasteur, 1890, p. 499.

AMERICAN.

Of the work done in investigating the diseases of swine that of Dr. F. S. Billings seems to have aroused much attention, chiefly because of the polemical attitude which he has assumed and the peculiar manner in which he has criticised the work of the Bureau of Animal Industry. His results are contained in a volume of 414 pages, published by the University of Nebraska in 1888, in connection with which institution the work was carried on.

In this volume Billings has given us very little opportunity to discover how much work was actually done in arriving at the various theories and conclusions contained in the volume. The statements of experiments are exceedingly meager and the bacteriological work very unsatisfactory.

It is evident that in work of this kind the discovery of the causes of disease is the fundamental problem, and all other problems must at the outset be subservient to this one. The bacteriological work is therefore the most important. A perusal of the report of Billings shows that he has contributed nothing whatever to the elucidation of this problem, while his obstructive attitude has confused and retarded the progress in the right direction in a marked degree, as the following statements conclusively prove:

He has assumed the position that there is but one infectious swine disease in the country, while the investigations of the Bureau of Animal Industry have maintained that there are at least two. The various criticisms which Billings has written of the work of the Bureau need not engage our attention here. They are interesting enough to be read in the original, but they all collapse in view of investigations published in the present volume. That the earlier investigations of the Bureau on swine plague were not absolutely demonstrative no one will deny. That they, however, pointed very directly to another disease is shown by pathological as well as bacteriological considerations—a severe pleuro-pneumonia associated with specific pathogenic bacteria easily differentiated from the bacteria causing hog cholera. The repeated occurrence of pathogenic bacteria in case after case of an outbreak could not well be overlooked or explained away by any pathological considerations. The duty of an investigator in another section of the country

118

would have been the careful investigation of his own territory as to the presence or absence of such diseases.

What is the nature of the bacteria described by Billings as the cause of the infectious swine diseases?

If we examine pages 103–116 of his report we shall find a brief account of the bacteria in question. If we examine his description and figures on page 104, and compare them with the swine-plague bacteria described in the foregoing pages, it will be observed that he has before him swine-plague bacteria and not hog-cholera bacilli. Of this he seems to be himself aware, for on page 111 he states that the hog-cholera bacteria, as described in the reports of the Bureau, have no existence, i. e., are fictitious.

Next, the growth on potato as observed by Billings is anything but that of swine-plague bacteria. It may be that of hog-cholera bacilli or of allied intestinal bacteria.

The remarks about gelatine cultures may apply to at least a dozen species of bacteria.

The observations about the movement of these bacteria is equally indefinite. It might apply very well to the Brownian motion of swine-plague bacteria, but it certainly does not describe the rapid motion across the microscopic field so characteristic of hog-cholera bacilli.*

This very incomplete description of the bacteria found by Billings leaves us, therefore, entirely in the dark. The form, mode of staining, and the motility, apply to swine-plague bacteria, the potato growth, perhaps, to hog-cholera bacilli. How can we reconcile this conflicting account? Bearing in mind the fact that in the organs of swine which have succumbed after extensive lesions of the lungs and large intestine, it is not an uncommon thing to find various forms of bacteria, *bacillus coli*, non-motile bacilli (also found in intestines), streptococci, gas-producing, spore-bearing bacilli, etc., either alone or associated with the real cause of the disease, we find ourselves unable to explain his discovery because we have no full and accurate report of investigations actually carried out. The description he gives may apply so far as we know to the bacteria found in one hog, in five hogs, or in a hundred hogs. It is obvious that the amount of conviction his statements carry depends entirely upon the number of animals to which such statements apply.

An explanation which covers the ground of the statements made by Billings fairly well is one which takes into account the mingling of two diseases. Since 1886 we have seen very few outbreaks of hog cholera not associated with swine plague. The investigations recorded in these pages show how frequently it may occur that a culture may contain

---

* In his report on the "corn-stalk" disease of cattle, page 186, Billings seems to have lost his doubts concerning the motility of the hog-cholera (swine-plague of Billings) bacillus. He there considers it by comparison "to possess most active movements."

both swine-plague and hog-cholera bacilli, or that the cultures from one animal contain only swine-plague bacteria, those from another of the same herd only hog-cholera bacilli. Again this report illustrates that we may be called upon to investigate an outbreak of swine disease in which hog cholera bacilli are demonstrable in every case, and in the succeeding one we may find only swine plague bacteria or a mixture of both germs.

It is highly probable that Billings had under observation now one germ, now another, and occasionally a mixture of both. With this explanation * in mind we may easily interpret the conflicting account of the bacteria given by Billings, especially if such statements are based on a small number of cases only. This explanation is the more plausible when we turn to the method used by Billings in obtaining pure cultures from animals. While, on the one hand, his cultivation methods were insufficient to determine accurately whether cultures contain more than one organism or not, he unwittingly assumes, on the other hand, first, that the bodies of diseased swine always contain only one kind of bacteria, and, second, that this kind is always the same. These assumptions anyone will recognize on reading page 103 of his report. How much information can be obtained by such deductive method of pursuing a most inductive branch of scientific investigation, the reader must be allowed to judge for himself. It may be noted, however, that such vicious methods furnish ample material for the attack upon work done by others.

Again, the examination of 5 hogs, or of 500 hogs, made over a limited territory with a uniform result, does not permit us to generalize negatively on the swine diseases of the remaining millions scattered over the whole country. This attitude is to be regretted the more

---

* A good illustration of the plausibility of this theory may be found on pp. 191-197 of the report of Billings, where the results of some inoculations of smaller animals are detailed. These may be tabulated.

One rabbit, inoculated with one-half cubic centimetre bouillon culture, subcutaneously, May 14, dies in 3 days.

One squirrel, inoculated with one-half cubic centimetre bouillon culture, subcutaneously, May 14, dies in 3 days.

One rabbit, inoculated with one-fourth cubic centimetre spleen emulsion of preceding rabbit, May 17, dies in 6 days.

One rabbit, inoculated with seven drops of blood, etc., from pig, subcutaneously, May 23, dies in 1½ days.

In this series the dose injected into the first rabbit is too large to bring out the differential characters of either hog-cholera or swine-plague germs. In the second rabbit death was probably due to hog-cholera bacilli. In the third it was certainly due to swine plague, since the most virulent hog-cholera bacilli would not destroy rabbits in the dose used, in less than 4 to 7 days. In the report on the corn-stalk disease already referred to (1889), Billings states that a certain germ could not "be that of swine plague (hog cholera) on account of its acute fatality." These later opinions are refreshing in being in the right direction as far as hog cholera is concerned.

when we reflect on the fact that the area of diseases is largely defined by geographical, climatic, and economic factors.

In his report, Billings throughout denies the existence of the hog-cholera bacillus first described in the Bureau Report for 1885. This denial may be found scattered throughout the report. Then we must assume that he considered the swine-plague bacteria as the cause of American swine diseases. This assumption is proven by his repeated, almost continuous, discussion of the European *Schweineseuche* and *Wildseuche*.* This continual dragging in of the work of European observers can only be understood by assuming a great similarity or a possible identity between the bacteria in the hands of Billings and those of *Schweineseuche*, as described by Löffler, Schütz, Hüppe, and others. On the other hand, the pathological appearances in these diseases differed so greatly that Billings found it necessary to occupy the greater part of his report in needlessly pointing out likenesses and differences.

Up to this point, then, it seems that Billings regarded his swine disease bacteria as the same morphologically with the German *Wildseuche*, and that he denied the existence of the hog-cholera bacillus. Meanwhile we studied his publications and made every effort to determine what bacteria he was studying, but without success, owing to the imperfect diagnosis given and the peculiar intermingling of the properties of the bacteria of hog cholera and swine plague found in his report.

In 1889, however, to our surprise, the commission appointed to throw some light on this matter found Billings in possession of real hog-cholera bacilli. Cultures were sent by him to Berlin, where a comparison of these cultures with some sent from this laboratory likewise demonstrated that his swine-plague bacteria were identical with the hog-cholera bacilli discovered in the Bureau laboratory in 1885,† nearly a year before Billings began his work in Nebraska.

All these facts make it highly probable that Billings had unwittingly studied at least several kinds of bacteria, among which the swine-plague bacteria of this report must have played an important part. In a recent article‡ on swine diseases, prepared under Hüppe, the author mentions the fact that Billings had sent cultures at two different times, which, though considered by the sender as identical, were really quite different. In another article Caneva,§ working under Hüppe in Fresenius' laboratory, endeavors to group the various bacteria producing swine disease, and in so doing separates the bacteria sent by Billings from hog-cholera bacilli, because the former were less actively motile, coagulated milk, and produced only local reaction after subcutaneous inoculation. They also failed to infect by feeding. While these bacteria may represent an attenuated variety of hog-cholera bacilli so far as their

---

* Compare also page 111 of this report.
† Frosch. Zeitschrift für Hygiene, IX, S. 235.
‡ Bunzl-Federn. Archiv f. Hygiene, XII, S. 198.
§ Centralblatt f. Bakteriologie, IX, S. 557.

virulence is concerned, none of the hog-cholera bacilli from numerous and widely separated localities examined in this laboratory ever produced coagulation of milk. Bacteria of the latter class are chiefly harmless intestinal parasites which closely resemble hog-cholera bacilli (*bacillus coli*) and which occasionally appear in cultures from cases of swine disease.

If, therefore, Billings had found the hog-cholera bacillus the question arises : Why did he fill up his report with pages of extracts, comment, and criticism of the European *Schweineseuche*, when the bacteria of hog cholera and *Schweineseuche* are wholly different, as every observer has admitted who has compared them ? Such discussions are not only useless, but misleading, when brought to bear upon the condition of things in our own country. In view of these facts the question also arises: Why did Billings so vehemently oppose the hog-cholera bacillus described in the Bureau Report of 1885, and found by us to exist in Nebraska before Billings entered upon his work there, if he had it under observation himself ? These contradictory positions can only be interpreted by the assumption made above, that Billings had at first one or several kinds of bacteria under observation differing from the true hog-cholera bacillus.

If we are nevertheless to conclude that Billings has finally settled upon hog-cholera bacilli as the cause of swine disease in Nebraska, any further comment on his work could only be taken up under hog cholera. The question whether there is another disease besides hog cholera is settled in the affirmative by the work reported in these pages. This second disease seems to be in fact the disease which Billings has had in mind in his controversies, and his criticisms of American and foreign work. Unfortunately, however, it has turned out that he has mistaken the disease, and now his opposition strangely enough has shifted towards the swine-plague bacteria, since an attitude of opposition towards hog cholera could no longer be maintained.*

It is clear to any unbiased reader that work which fails to grasp any positive truth, and is continually shifting its base to avoid the necessary consequences of serious errors, and which goes beyond its confines not only to criticise, but to discredit in every manner possible the work of other observers, can not be seriously taken into consideration as advancing in the least degree our positive knowledge.

During the years 1887–'89, Professor Welch, in conjunction with A. W. Clement, V. S., and F. L. Russell, V. S., investigated a number of outbreaks of swine disease in the neighborhood of Baltimore, Md. In

---

* A simple statement of the position of Billings would read thus :

1. Opposition to hog-cholera bacteria discovered in Bureau laboratory in 1885. Evidence in his writings all points to swine plague,

2. Discovery by commission and German observers that his germ and the hog-cholera bacillus are identical ; hence,

3. Opposition to swine-plague bacteria.

a preliminary report published in the Johns Hopkins Hospital Bulletin, December, 1889, Welch gives a brief summary of the results obtained up to that date. These investigators encountered in some herds only hog-cholera bacilli, in others only swine-plague bacteria, and in still others both kinds of bacteria. They have not been able to fix upon any anatomical differences between the herds in which hog-cholera bacilli were found and those in which only swine-plague bacteria were detected, for in all cases intestinal lesions were present. The description of the two kinds of bacteria agrees in every respect with that published in the various reports of the Bureau of Animal Industry since 1885. With both the authors were able to produce disease in swine, intestinal lesions with hog-cholera bacilli and lung lesions, associated with inflammation of the serous membranes, with swine-plague bacteria.

While the results agree in every particular with those obtained by us, Welch expresses himself with caution concerning the rôle of swine-plague bacteria, because in the herds studied none were free from intestinal lesions. He suggests the possibility of overlooking hog-cholera bacilli because they may remain limited to the intestinal tract, a possibility to which we have called attention in the report for 1887–'88, and in these pages. On the other hand the facts that pneumonia may be produced by swine-plague bacteria, and that a swine disease exists in Germany in which pneumonia without intestinal lesions is associated with swine-plague bacteria, "suggest that this organism is also the cause of a similar affection in this country."

In 1889 and 1890 Dr. J. A. Jeffries[*] made bacteriological observations in several outbreaks of swine disease of an infectious character. In one pig were found a large spore-bearing bacillus, a short bacillus, and swine-plague bacteria.[†] The short bacilli Jeffries found non-pathogenic, while the third form, the swine-plague bacteria, he considers the cause of the disease.

The description of the pathological appearances of the diseased pigs and of the swine-plague bacteria found by him, taken together with the inoculation experiments and the absence of hog-cholera bacilli, make it pretty certain that the disease was identical with that described in these pages.

Through the kindness of Dr. Jeffries a culture of the swine-plague bacteria was sent to the laboratory, where a comparative study of the morphological and pathogenic characters showed them to be an attenuated variety of swine-plague bacteria, not distinguishable from those described in these pages. A few of the inoculations on rabbits

---

[*] Etiology of two outbreaks of disease among hogs. The Journal of Comp. Medicine, December, 1890.

[†] The spore-bearing bacillus I have found in many outbreaks as the result of post-mortem growth. In some pigs sections from every organ show these long wavy filaments filling up the capilaries and penetrating the tissue in all directions. See also page 20 of this report.

are given to illustrate the variety of lesions which these bacteria may produce.

The culture was first tested by plate cultures and two rabbits inoculated from a bouillon sub-culture.

June 3, 1890. One rabbit received one-eighth cubic centimetre subcutaneously, and one one-eighth cubic centimetre into an ear vein.

In both the temperature was between 105° and 106° F. on the sixth day. On the tenth day the first had recovered; the second, unable to move hind limbs, was chloroformed. The only lesions observed were two abscesses, one on the right tarsus, the other on left elbow joint, both communicating with the joint cavity.

The culture was thus considerably attenuated. Nothing more was done till December, when the same culture, passed through a series of agar tubes meanwhile, was used, because attenuated, for some preliminary immunity experiments on rabbits. The virulence of the bacteria was much greater now, as the following inoculations prove. Attention is called to the great variation in the lesions produced, and to the pneumonia in No. 5.

December 8, 1890. Two rabbits (Nos. 1, 2) received subcutaneously one-eighth cubic centimetre bouillon culture, and two (Nos. 3, 4) the same dose into an ear vein.

No. 1 dies in 7 days with extensive purulent infiltration of the subcutis over abdomen and thorax, purulent peritonitis and pleuritis.

No. 2 survives.

No. 3 very sick on the fifth day and chloroformed. A subcutaneous purulent infiltration extends from place of injection on the ear over the greater part of face. No other lesions observed.

No. 4. Temperature on third day 105.5° F. Dies on the fifth day with peritonitis and pleuritis. The exudate stretches in the form of delicate grayish viscid threads between coils of intestine when these are lifted up, and from chest-wall to pleura of lungs. The exudate a mixture of leucocytes and immense numbers of bacteria.

As these experiments were designed to find a dose which would not prove fatal they were repeated.

December 16. Two rabbits (Nos. 5, 6) receive a subcutaneous injection of one-eight-hundredth cubic centimetre bouillon culture, and two (Nos. 7, 8) an intravenous injection of the same dose.

In all cases the culture was diluted and one-fourth cubic centimetre of the dilution injected.

No. 5 dies in 9 days. Extensive purulent infiltration of subcutis over abdomen and part of thorax. Exudative peritonitis absent. Purulent pleuritis with exudate especially abundant on right lung and chest-wall. *Hepatization of the two small anterior and portion of principal lobe of same side.* The hepatized lobes in part dark red and pale red, firm and enlarged. Epicardium covered with a membranous exudate. In the exudate numerous bacteria showing the polar stain.

No. 6. Temperature on second day 105° F. Dies on sixth day. Extensive purulent and sanguinolent infiltration of the subcutis as in preceding case. Spleen barely enlarged, somewhat darker than normal. No peritonitis or pleuritis.

No. 7. Temperature 105.2° F. on second day. Dies on thirteenth day. Extensive subcutaneous infiltration as in preceding case. Straw-colored, elastic membranous exudate on liver, spleen, and cæcum, and between coils of large intestine, matting the various organs together. The exudate is easily pulled away and consists of fatty pus cells and immense numbers of swine-plague bacteria. In pleural sacs some serous exudate. Pale red hepatization of a small portion of the small anterior lobes of right lung. Spleen moderately enlarged, dark. Peyer's patches pigmented.

In November, 1890, Prof. T. J. Burrill, of Illinois University, sent two agar cultures of bacteria obtained from an outbreak of swine discase in Illinois. Both cultures were carefully examined and found to contain only swine-plague bacteria. To test the pathogenic character two rabbits were inoculated. These inoculations prove the essential identity of these bacteria with swine-plague bacteria.

November 7. From a bouillon culture, 1 day old, one-tenth cubic centimetre was injected under the skin of a large rabbit.

November 11. Rabbit found dead this morning. At point of inoculation purulent thickening of the subcutis. Cæcum, colon and rectum sprinkled with hemorrhages. Slight viscid exudate on cæcum which contains large numbers of the injected bacteria. Liver and kidneys pale. Spleen barely enlarged, dark in color. The agar and bouillon cultures from blood, spleen, and abdominal exudate contain only the inoculated bacteria.

November 13. A rabbit received subcutaneously on side of abdomen one-tenth cubic centimetre bouillon culture prepared from the second original culture. Rabbit dies in 40 hours. A very slight infiltration at the place of inoculation. Spleen slightly enlarged and engorged. In it a considerable number of swine-plague bacteria showing the polar stain. An agar culture from the spleen contains these bacteria only.

GERMAN.

Of great interest and importance to us are the German investigations of swine plague (*Schweineseuche*), because this disease does not appear to be complicated with any other disease, as is the case in our own country where swine plague and hog cholera are so frequently associated with each other.

Probably the first investigation in which swine-plague bacteria were observed is that of Löffler.* Only one pig was encountered in the course of a series of investigations on the continental disease of swine, known as *rouget* and *Rothlauf*, in which these bacteria were found. In this case the lesions are given as follows :

The skin of abdomen, sexual organs, and neck of a livid red ; enormous œdema of the (sub) cutis of the neck, extending backwards between forelimbs. Pharynx reddened and swollen. Laryngeal and tracheal mucosa intensely dark red. Lungs but slightly affected ; on the right some dark red regions, containing but little air. Nothing abnormal about the heart. Cloudy swelling of liver and kidneys. Mucosa of stomach and upper portion of duodenum intensely reddened. Remainder of intestines unchanged. Mesenteric glands not enlarged. Spleen rather large, dark bluish red, quite firm.

The bacteria obtained from this case resembled those of rabbit septicæmia.† They killed inoculated rabbits and mice in 24 hours. Guinea-

* Arbeiten a. d. kaiserlichen Gesundheitsamte, 1 (1885), S. 51.

† The term rabbit septicæmia originated in a series of experiments on rabbits by Gaffky, in 1881 (*Mitthetlungen a. d. kaiserl. Gesundheitsamte* 1, 1881, S. 102), who produced in these animals by the injection of polluted water a rapidly fatal disease caused by bacteria closely resembling those of swine plague and some other animal diseases. (See p. 141 of this report.) The writer found bacteria causing septic diseases in rabbits, probably identical with these in 1886 (Journ. Comp. Med. and Surgery, Jan., 1887). Prior to Gaffky's work, R. Koch and Davaine had been experimenting with a similar disease in rabbits.

pigs lived from 2 to 3 days after inoculation. In all there was extensive sero-sanguinolent infiltration of the subcutis, starting from the place of inoculation and extending in some over thorax and abdomen. In one the intestines were covered with a sanguinolent and fibrinous exudate. Of three pigs inoculated (probably with minimum doses) one died in 2 days with the following lesions :

Skin of abdomen bluish red ; enormous œdema of skin ; lungs hypostatic ; mucosa of stomach deeply reddened ; spleen unchanged ; kidneys parenchymatous ; mesenteric glands not swollen.

This first reported case is interesting in that the lungs were not the seat of disease.

The disease of *Schweineseuche* was established in a more definite manner by Schütz in investigations carried on in 1885.*

Since then it has been generally recognized as a disease distinct from *Rothlauf*. The material with which Schütz worked at this time consisted of—

1. The stomachs, spleens, and livers of three pigs, more or less decomposed, June 15, 1885.

2. The stomachs and spleens of several pigs, August 27, from Putlitz.

3. The trunks of two pigs, November 19, from Putlitz.

4. Two entire pigs, December 13, from Putlitz.

In a footnote the author states that several additional cases of this disease had come under his observation subsequently.

In the four pigs of which Schütz was enabled to examine the viscera, all were affected with more or less hepatization of the lungs associated with pleuritis, more rarely with pericarditis. In one of these four cases there was found, in addition to lung disease, peculiar caseous degeneration of the joints of the limbs, involving the bones and surrounding muscles. The various lymphatic glands were greatly enlarged and contained cavities filled with grayish yellow, semi-liquid masses.† In the diseased lungs were disseminated yellowish necrotic foci varying in size. From all the cases examined bacteria were obtained which were evidently the same. Schütz described them as follows:

When stained with gentian violet they show in their central portions an unstained region surrounded by a layer stained blue. The thickness of this layer is greater at the poles, so that the extremities appear more deeply stained than the sides. When deeply stained they appear uniformly blue. As these organisms stand between micrococci and bacilli, they may be called bacteria. They are 1.2 μ long and 0.4 μ to 0.5 μ broad. They multiply in the following manner: They become twice as long as broad ; show distinctly rounded extremities, and stain like the organisms of rabbit septicæmia and fowl cholera, so that between the deeply stained ends about one-half or a third of the entire length remains unstained. Careful examination shows, however, that the colored end pieces are connected with each other by a fine line which passes from one to another on each side. The end pieces then separate and the median portion disappears. The former are at first spherical, but very soon assume an oval form. Hence from every organism two new individuals arise by division, in

---

* Loc. cit., 1 (1885), pp. 376-413.
† Compare cases on p. 75 of this report.

which by careful staining the uncolored central portion is easily distinguished from the colored periphery. If the process of multiplication is very rapid, as in pigs and rabbits, the organisms do not attain the size given above, but divide before the unstained median piece becomes distinctly visible. Under these circumstances the organisms of the succeeding generations are smaller, only one-half as large as, or even smaller than, those which have resulted from the slow division of the bacteria. The younger generations are frequently extraordinarily small, plainly oval, however, and staining uniformly in gentian violet. They do not execute any spontaneous movements.

These bacteria were fatal to mice, rabbits, and guinea-pigs. Pigeons succumbed to large doses. Fowls and rats were not susceptible. A comparison of the virulence of these bacteria obtained from the organs of animals at the four different times indicated above, the last three being from the same locality, shows that it varied slightly:

1. Rabbits died in 2 days, mice in 1 and 2 days after inoculation. One pig 4 to 5 months old, which had received subcutaneously the contents of two Pravaz syringes of bouillon culture subcutaneously, died in 24 hours. A second pig treated in the same way died in 48 hours.

2. Mice died in 24 hours.

3. Mice died in 24 hours; rabbits in 2 to 3 days; guinea-pigs in 4 to 8 days.

4. Mice and rabbits died in 24 hours; guinea-pigs in 2 to 5 days; one pig, which had received a syringeful of a bouillon culture into each lung, died in 2 to 3 days.

The lesions in guinea-pigs, rabbits, and mice are, in the main, those obtained with the bacteria of our swine plague. The above table indicates that the bacteria from the last lot were the most virulent, as they were fatal to rabbits in 24 hours.

Through the kindness of Prof. W. H. Welch, of the Johns Hopkins University, a culture of the German swine-plague germ was obtained in 1889, and a second fresh culture in 1890, both from the Berlin Hygienic Institute. Both were compared with the American varieties of the swine-plague bacteria. While the bacteria in both cultures were identical in form and biological characters with the swine-plague bacteria of this country, their pathogenic properties varied somewhat, as will be seen from the following experiments:

June 18, 1889. From the original agar tube agar roll-cultures were made, and from 2 colonies peptone-bouillon cultures prepared.

June 25. Two rabbits received subcutaneously one-eighth and one-sixteenth cubic centimetre, respectively, 2 mice each about one-sixteenth cubic centimetre.

June 26. One mouse dead this morning. Large, dark spleen, fatty liver. No bacteria detected in cover-glass preparations from blood, spleen, or liver. On the following day a few colonies had appeared in the agar culture from spleen.

June 27. Second mouse dead. Spleen slightly enlarged, liver fatty. In spleen and blood very many swine-plague bacteria, exhibiting the polar stain.

Neither of the rabbits died. The one which had received one-eighth cubic centimetre had a temperature of 106.4° F. on the third day. The temperature of the other was not taken. Both were killed 18 days after the inoculation. The first rabbit considerably emaciated. On the inoculated thigh an abscess as large as a hen's egg, discharging from an opening thick pus. The suppuration had extended to the abdomen, where over 6 square inches of the subcutis was infiltrated with pus and firmly adherent to abdominal muscles. In the second rabbit a small abscess as large as a marble on inoculated thigh. Spleen considerably enlarged. Peyer's patches swollen.

The attenuated condition of these bacteria discouraged any attempts to determine their effect upon swine.

The second culture received in 1890 proved to be a far more virulent type, since inoculation of rabbits was invariably fatal within 20 hours. These bacteria did not differ therefore in this respect from those obtained from outbreaks VII and IX. At the same time their virulence was still greater, as will appear from the following trials upon swine:

January 9, 1891. Pig No. 435, black male, mixed grade, 4½ months old, received subcutaneously into each thigh 2½ cubic centimetres of a peptone-bouillon culture, or 5 cubic centimetres in all. Dies just 24 hours later.

Pig No. 437, black and white female, 4½ months old, one-half cubic centimetre of the same culture injected into a vein of leg. Animal struggled so that the quantity injected may have been more or less than one-half cubic centimetre. Dead in 36 hours.

In both animals there was considerable necrosis of the skin and œdema of the subcutis where the inoculation was made. These animals were not examined until postmortem changes had appeared, owing to other work, so that the autopsy notes are omitted. The following two cases are of interest in that the quantity of culture liquid injected was smaller.

February 11. No. 460, black and sandy female, 3½ months old, weight 60 pounds, inoculated subcutaneously with 1 cubic centimetre of a bouillon culture 24 hours old, one-half injected into each thigh.

No. 461, animal of the same kind, inoculated in the same manner with 3 cubic centimetres.

Both were found unable to rise on the following morning and died at 8 p. m., about 28 hours after inoculation.

Autopsy early next morning. No. 461 in good condition. General blush of skin on ventral aspect of body and limbs. Slight reddening of subcutaneous fat. On both inoculated thighs the connective tissue reddened and all minute vessels injected. On the right the connective tissue has also a glistening œdematous appearance which extends upon abdomen as far as umbilicus.

Considerable blood extravasation on pericardium and on epicardium along the base of the heart, interventricular grooves and left ventricle. Veins on the surface of the heart distended. In right side a very dark, soft clot imbedded in thick, tarry blood. Lungs normal. Intestines have a uniformly reddened appearance from the outside. Stomach about half full of food. Mucosa of fundus hyperæmic. Mucosa of small intestine normal. In large intestine much dry feces. Mucosa of cæcum and colon of a wine-red color. Liver somewhat flabby. The surface has a mottled appearance due to the varying hyperæmia of the lobules. In the gall-bladder a firm body which almost fills it out and has the appearance of beeswax. The body is readily crushed with the fingers. Surrounding this body is a yellowish-white, pasty mass. Kidneys hyperæmic.

In cover-glass preparations of blood and kidneys many bacteria exhibiting the polar stain; in liver, only a few. On inclined agar traces from blood, kidney, and liver transferred with wire gave rise to a very dense growth of swine-plague bacteria.

No. 460. From the cut ends of the subcutaneous veins thick, dark blood exudes. Skin and subcutis as in No. 461, the vascular injection on the thighs more pronounced.

Intestines and stomach appear much reddened from the outside. On the abdominal walls and coils of large intestine are little lumps of yellowish-white exudate. A few coils of the small intestine where they touch each other show bands of petechiæ under serosa.

Condition of heart as in 461. The right ventral lobe of lungs fastened to pericardium by two old adhesions. Some subpleural hemorrhages on principal lobes. Slight roughening of pleura over the ventral half of both lungs. Parenchyma normal.

Mucosa of fundus of stomach over an area 6 inches in diameter much reddened, the hyperæmia extending to submucosa. Small intestine contains occasional patches of congested mucosa. In the large intestine hyperæmia slight compared with 461. Kidneys and liver as in 461.

In the blood many swine-plague bacteria. Cultures therefrom and from kidneys confirmatory.

It should be noted that while these German swine-plague bacteria were fatal after subcutaneous inoculation, the only cultures of the American variety which killed swine after such inoculation were from outbreaks I, II, and IX. Even the bacteria from the latter outbreak, virulent as they were, failed in this respect in all but one case.*

Additional investigations concerning *Schweineseuche* were made by Bleisch and Fiedelert in 1888–'89. The disease appeared in September on a dairy farm, evidently introduced by recently purchased animals. It spread among the swine in several different stables, even among those which did not come in direct contact with the purchased animals, but were simply placed in the pens evacuated by the latter. Even after the disinfection of two stables the animals put into them contracted the disease. The investigations were continued until February, 1889, and in all fifty-two animals were examined. The disease had been comparatively mild and chronic; none of the infected died, and the lesions were observed in the slaughtered animals.

During life the symptoms consisted of coughing, which increased in severity when fresh air entered the stables, difficult breathing, loss of appetite, and emaciation. The temperature fluctuated between 102° and 105° F. The lesions observed in the butchered animals were in the main limited to the thoracic organs. The anterior (or cephalic) lobes were involved in grayish-red hepatization, which in more advanced cases invaded the middle and posterior (ventral and principal) lobes, the latter only in isolated regions. The bronchus always formed the central point of the hepatization. As the disease progressed partial caseation of the hepatized tissue and of the bronchial glands took place. The caseous masses did not contain tubercle bacilli. Pleuritis was found only in advanced cases, pericarditis still more rarely.

The authors found in practically all cases bacteria, which they identify with the bacteria of *Schweineseuche* or swine plague. They were obtained both by inoculating rabbits and fowls with particles of lung tissue, and more rarely on plate cultures from the lung tissue directly. The bacteria obtained are so far as the description goes identical morphologically with the bacteria found by Löffler, Schütz, and others. Their effect on rabbits differs in some respects from the disease produced by the swine-plague bacteria proper. The disease

---

* See p. 74.
† Zeitschrift f. Hygiene, VI (1889), S. 401–452.,

may last from 2 to 13 days. An examination of the text shows that some rabbits lived longer, one 27 days. At the point of inoculation there is more or less subcutaneous purulent infiltration with purulent lymphangitis. The liver contained in many cases embolic foci, which are described (p. 406) as partly branched, partly yellowish-white spots found on the surface and on section, and consisting of a pasty mass. A careful examination of the text fails to convince me that these embolic foci had any connection with the disease. Their description suggests very strongly cysts of *coccidium oviforme* in various stages of enlargement. Thus a rabbit which died in 24 hours from an intrathoracic inoculation of these swine-plague bacteria showed at the autopsy "the liver very large, the surface studded with numerous, projecting, yellowish nodules as large as pease, similarly the cut surface." The coccidia, if such they may have been, are easily overlooked in cover-glass preparations, for the method of preparation and staining destroys them pretty thoroughly. Moreover, these embolic foci do not, according to their report, appear with any regularity in the inoculated animals.

Another point to be noted in the lesions of inoculated rabbits is the absence of exudative peritonitis or pleuritis, which is a very constant lesion in rabbits inoculated with the American races of swine plague bacteria when they live more than two days after inoculation.

While swine-plague bacteria have little or no effect upon fowls unless large doses are injected into the muscles, the variety under discussion was virulent enough to prove fatal to almost all fowls inoculated. Death occurred from one day to several weeks after inoculation. In one of these "liver emboli" were observed. One pig which had received a Pravaz syringe full of bouillon culture of these bacteria died in ten hours. The lesions observed were pleuritic effusion, hepatization of almost the whole of the right lung. In the left there were isolated masses of hepatization. A second pig which received a subcutaneous injection had a slightly elevated temperature for several days, but it recovered subsequently.

In several diseased pigs there were found ulcers on the surface of the body. In one a series of caseous cysts, starting from the castration wound and extending along the subperitoneal tissue as far as the umbilicus, was observed. The relation of these lesions to the disease is, of course, not determinable.

The authors, after having determined the cause of the disease, endeavored to find out how the bacteria are transmitted from one animal to another. The swine examined belonged to one farm and were fed chiefly with sour whey. In this whey, taken from the troughs, bacteria probably identical with the disease germs were found on two different occasions. Further investigation revealed the fact that while fresh milk is not a good soil for these bacteria, sour whey is very favorable to their multiplication. The authors therefore explain the transmission of the disease by assuming that in the common feeding trough the bacteria are mixed with the milk.* Some of this accidentally getting into the air passages during feeding introduces the disease germs.

This brief review of the investigations indicates that while the bacteria found by Bleisch and Fiedeler are not hog-cholera bacilli and in general the same as swine-plague bacteria, there are some minor but constant differences to which attention has been called. It is a curious fact that these bacteria were attenuated with reference to rabbits, but still fatal to fowls. Towards the varieties described in this report, the fowl, among smaller animals, manifested the greatest and the rabbit the least power of resistance. It is not improbable that attenuation, as we understand the term, may imply a decrease of virulence as to one species and at the same time an increase as to others. It is not unreasonable to assume that a variety of swine-plague bacteria, apparently attenuated so far as rabbits are concerned, may still possess virulent properties as regards more refractory animals. The properties which make it feeble to rabbits and virulent to swine, for instance, may prove different from each other. It is only by such assumption that we can understand the action of swine-plague bacteria from outbreak I and II, which, though attenuated with reference to rabbits, were fatal to swine.

A very interesting communication on the subject of swine diseases in Germany was recently made by F. Peters, of Schwerin.† During the winter of 1887–'88 this author examined cases of a disease among swine which strongly suggests hog cholera. The disease is described in brief as follows:

Soon after the sucking period is completed, the young pigs lose their desire for food and become emaciated. Cough, increased respiration, paleness of the mucous membranes, and diarrhea are also observed. Towards the fatal close of the disease, which lasts from 3 to 6 weeks, the skin of the ears, the neck, and chest becomes reddened. The greater number of those attacked die. In four cases, the large intestine only was affected. The description given corresponds closely with the various forms of necrosis, softening and induration commonly called ulcers in hog cholera for the sake of simplicity. In a fifth case, in addition to the intestinal changes, there were found recent hepatization of the left lung and exudative pleuritis.

The author made some bacteriological examinations, but they were not thorough enough to furnish any reliable information concerning the character of the bacteria found. While he maintains that they are swine-plague bacteria the description he gives would equally apply, as far as it goes, to hog-cholera bacilli. As the territory of Schwerin is not very far from Denmark, the scene of swine pest (hog cholera) during 1887 and since that time, it is not improbable that this region is slowly being invaded by two diseases, the Schweineseuche, or swine plague,

* The American varieties of swine-plague bacteria refuse to grow in acid media.
† Die Schweineseuche. Archiv f. Thierheilkunde, 1890, XVI, S. 64.

found by Schütz in 1885, and the swine pest, or hog cholera, first noticed in Sweden and Denmark in 1887.* This possibility has been suggested recently by Bnuzl-Federn† in an article devoted to swine-plague and closely related bacteria. The problem of infectious swine diseases would then enter the phase in which the investigations of this Bureau found it as far back as 1886 in this country, in which a mixture of two diseases is encountered more frequently than either disease by itself.

In this summary some articles of minor importance have remained unnoticed. Likewise the observations of Roloff‡ on caseous changes in the intestines of young pigs have been passed by because they give us no information as to the nature of the bacteria involved in the disease. The views of some that his cases were swine plague, of others that they may have been hog cholera, are purely hypothetical.

* See Special Report on Hog Cholera, 1889, p. 181.
† Archiv f. Hygiene, XII, 1891, S. 198.
‡ Die Schwindsucht, fettige Degeneration, Scrophulose und Tuberkulose bei Schweinen. Berlin, 1875.

SOME PRACTICAL OBSERVATIONS ON THE PREVENTION OF SWINE PLAGUE.

1. CONDITIONS WHICH MAY FAVOR AND OPPOSE OUTBREAKS OF SWINE DISEASE.

The factors which enter into the production of outbreaks of swine plague may be divided for convenience into two classes, those pertaining to the animal itself, and which make it more or less susceptible or insusceptible to the specific bacteria, and those which relate to the bacteria.

The conditions which make animals more susceptible to infection are as varied as the conditions which reduce their vitality. The importance of rearing and keeping animals in such a manner as to produce and maintain a healthy action of the various functions of the body has not been insisted upon with as much emphasis as it deserves, owing to the somewhat overshadowing influence which the study of pathogenic bacteria has exerted upon all minds. It is evident, however, that veterinary hygiene has much to do with the decline of large epizoötics, not only by keeping away the germs of disease, but by enabling the animal body to resist their attacks. Of those conditions of swine which invite disease very little is as yet positively known, and we simply call attention to a few to arouse the interest of those who are in position to make observations.

There have been indications during the course of experiments at the Bureau Station that the breed may have some influence in predisposing to infection. As an illustration we may cite an experiment in vaccination of swine against hog cholera carried on at the Station in 1889–'90.* The vaccination, which consisted in subcutaneous inoculation of culture liquid, seems to have had no effect; for, when the time for exposure came, practically all pigs from one lot succumbed and all from another lot survived. The latter were Essex grades reared in pens; the former, grades of mixed Jersey Reds and Chester Whites not raised in pens. While it is impossible to give any facts as to the relative resistance of different breeds to swine diseases, it is a subject which should receive the due consideration of swine-breeders, especially in those States where swine diseases are more or less stationary.

Age is another important element. We have found a decided difference in the susceptibility to both hog cholera and swine plague in favor of older swine. This element of age is familiar to all with reference to

* Report of the Secretary of Agriculture for 1890, p. 110.

certain human maladies. such as scarlet fever, measles, diphtheria, and some other diseases which preferably attack the young.

Feeding is perhaps the most important factor in predisposing swine to disease. The assimilation of large quantities of food and its conversion into fat seems to be the one essential function of swine. This goes on to such a degree as to lead to pathological conditions after a time. Not only the ingestion of large quantities of food, but of one kind for a long time, is in itself opposed to the habits of such omnivorous animals. Besides overfeeding upon one kind of food we have the uncleanly surroundings in which swine are apt to be kept contributing materially to a reduction of vitality.

In addition to the unhealthful modes of existence to which swine are subjected, and partly springing from them, are certain pathological conditions induced by parasites of different kinds. The life history of some of the more important parasites infesting swine is still to be elucidated. As a rule, we have found in our post-mortem examinations a larger number and variety of internal parasites in those herds which have been allowed to run freely than in those brought up in pens. The opportunities for infection seem to be much greater in the former case than in the latter.

As to the damage done by parasites it is difficult to form an accurate estimate from ordinary observation. Obvious damage may be done in the air passages by lung worms (*strongylus paradoxus*) and in the small intestine by *ascaris* and *echinorhynchus*. The lung worms may be met with in all seasons of the year in swine up to 3 months old. They invariably inhabit the terminal portion of the two large bronchi of the principal lobes. Here there is generally a partial or total occlusion of the bronchus for 1 or 2 inches from the caudal border of the lobe, due to the lung worms and the enveloping mucus. In some cases the occlusion is followed by collapse and broncho-pneumonia of the lobes supplied by the bronchus and its branches. The hepatized lung tissue assumes a bright or pale red color. When the lung worms are very abundant larger branches of the same bronchus become filled with these parasites and the broncho-pneumonia may extend over a greater portion of the principal lobes. That lungs in this condition are more susceptible to the invasion of swine-plague bacteria will be generally admitted. The bronchitis begun where the lung worms mature may extend after a time into the other air tubes. In outbreaks VII and VIII lung worms were found in almost every animal examined.

Another question arises with reference to lung worms as the possible bearers of the bacteria into the lungs. This will not be answered until more is known of the life history of these parasites. Meanwhile the evidence would hardly support the opinion that they may introduce the virus. The pneumonia usually begins in the small ventral lobes and travels from them while the lung worms begin their injurious work in the principal lobes farthest removed from the ventral lobes. All that can be said is that they may make the lungs more susceptible to the disease.

In the intestines *ascarides* are not infrequently found extending into the common bile duct from the duodenum. Some even enter the gall-bladder, while others imbed themselves in the ducts coming from the various lobes of the liver and completely obstruct the flow of bile. The *echinorhynchus* is well known as attaching itself to the mucous membrane of the small intestine, and producing ulcerous depressions simulating those of hog cholera.

That there may be other predisposing causes at certain seasons of the year, such as obscure malarial diseases due to protozoa, the invasion of the muscular system by psorospermia, trichinae, etc., need simply to be mentioned, since no positive evidence is at hand.

The important factor in the production of swine plague (and hog cholera as well) bearing on the bacteria is their virulence. We have seen in the chapter on the pathogenic action of swine-plague bacteria that their virulence or disease-producing power is subject to considerable variation, and that they may be very virulent as obtained from one outbreak and much less so from another. It may be laid down as a general rule that the more virulent the bacteria the more severe the resulting epizoötic, and the greater the mortality. While a more attenuated variety of bacteria may spare the older and more hardy animals of a herd, these will succumb to a more virulent variety. Just here the facts presented under the preceding head show their importance. Attenuated or weaker varieties of swine plague may attack the young and the badly kept swine, those infested with parasites and those of poorer breeds, while the stronger may not become diseased. This may explain also why some herds of swine are destroyed and neighboring ones escape, although both may have had the same opportunities of infection.

The appearance of an epizoötic depends thus upon the condition of the herd and the relative virulence of the bacteria. While there are bacteria whose virulence is sufficient to sweep away every obstacle, we are convinced also that much disease due to attenuated bacteria could be counteracted by a more hygienic breeding and rearing of swine.

2. THE DISTRIBUTION AND TRANSMISSION OF SWINE-PLAGUE BACTERIA.

We have seen in preceding pages that besides the particular herd in which swine plague exists as an epizoötic, bacteria not distinguishable from those of swine plague are found widely distributed in the air passages of healthy swine, and of other domesticated animals, such as cattle, dogs, and cats. Are these bacteria capable of producing disease in swine at any time and therefore a continual source of danger, or are they harmless? This question can not be answered definitely in the present state of knowledge on this subject. As a rule, the bacteria found in healthy animals belong to more or less attenuated varieties and are most likely incapable of producing disease excepting when the condition of the animals is very poor. Sporadic disease in such debilitated animals is not contagious and does not spread to other animals of the herd unless all are in equally reduced condition. It may happen, however, that such swine plague bacteria, which live in the air passages of older swine as survivals of former exposure and disease, may become dangerous to young pigs. Of this possibility outbreak VIII may serve as an illustration. The litter of young pigs died of swine plague caused by a considerably attenuated variety of bacteria, such as may be found in apparently healthy swine of greater age. It should also be remembered that even older swine, which have been through the fattening process and are, commercially speaking, in the best condition, are really in an abnormal or a pathological state, and, therefore, may be more or less susceptible to infection.

We have shown that there are herds of swine from certain farms entirely free from pathogenic bacteria, and the question arises, What is the source of those swine plague-like bacteria found in the upper air passages of many herds? I am inclined to think that they are transmitted from older swine to younger ones, producing disease only under aggravated conditions in isolated cases, which disease does not spread to other animals.

When we come to the more virulent varieties, those for example, which destroy rabbits within 16 hours after inoculation, the case is entirely different. Their presence is probably never manifested excepting by disease, and it is against the introduction of these bacteria that the swine-breeder must protect himself. Such bacteria always come from some outbreak of disease directly or indirectly. Let us consider briefly through what agencies such bacteria may be introduced into a herd.

In the report on hog cholera and in the present volume it was pointed out that swine-plague bacteria are far less hardy than hog-cholera bacilli.* The former perish rapidly in water and in liquids unsuited to their multiplication. They survive drying for a few days only. In general, they speedily disappear after they have left the body of diseased swine, and it is highly doubtful whether they would survive a month in the soil or in pens. Such agencies as streams, manure, etc., which may distribute hog-cholera bacilli over considerable distances are of restricted importance in swine plague. The chief danger lies in contact with diseased or infected swine. Intermediate carriers of infection can only act for a short time, while swine may harbor disease germs for months in localized inflammations, such as abscesses under the skin and in the joints, and it is possible that they may vegetate on the mucous membranes of the air passages much longer.

---

* See table, p. 92.

Swine must thus be regarded as the chief vehicle of infection. This may be conveyed directly from diseased to healthy animals; it may be conveyed by those which have passed through the disease, and hence by older to younger swine. It is safe to assume that any swine which have at any time been exposed to swine plague (or hog cholera) are liable to convey the disease, because we do not know when the specific disease germs leave the body.

Other sources of danger are railroads leaving fresh manure in different places, the vicinity of slaughter-houses, rendering establishments, or any places where the viscera of swine may be scattered or where numbers of living swine are temporarily housed. If we bear in mind the wide distribution of infectious swine diseases it is easy to believe that in any large herd of swine collected from different localities there are always some diseased or infected. It is essential, therefore, in guarding against disease, to look with suspicion upon all swine the history of which is not known to some extent at least.

There is a practice current in some parts of the country, and well illustrated by the history of outbreak IX, of gathering together herds of young pigs from various localities through the intervention of dealers. In regions where swine diseases are prevalent much of the time, and where the virus never dies out, this is a specially dangerous practice. While swine may not be visibly diseased, or may simply appear somewhat unthrifty, they still may carry the seeds of a virulent outbreak within them which need a little time to gain the required momentum. The mild character of a disease in any one animal is no evidence of the character of the germ. For this mildness may be due to a very virulent germ acting upon a highly insusceptible animal and causing a more prolonged chronic disease. In fact, these partly insusceptible animals are the most likely to appear in the markets because they are the remnants of herds destroyed by disease. We have frequently been able to demonstrate by experimental inoculations the general accuracy of these statements. Thus bacteria obtained from inoculated cases which had assumed a more chronic course had not lost any of their virulence. In experiments bearing on vaccination we have been able to increase the insusceptibility of rabbits and guinea-pigs so that virulent bacteria produced only a mild form of the disease, prolonged from days to weeks and even months. Yet the bacteria cultivated from such cases and injected into animals not vaccinated showed no loss of virulence. Again, we have found swine-plague bacteria in apparently healthy swine inoculated two months previously, and in case of hog cholera we have found the bacilli in the organs of swine 6 to 7 months after apparently unsuccessful inoculations. These bacteria possessed the original virulence.

The question has frequently arisen in the course of these investigations whether the bacteria are ever introduced into herds in the food. This involves another question, whether hog-cholera or swine-plague bacteria do exist independently of diseased or healthy animals. As to both kinds of disease germs there is no evidence that they live outside of the animal organism, except temporarily, and that if the food happens to be infected the infection has come from animals directly or indirectly, and that it is simply a question of time whether such infection is still in a living condition or not. Food, however, may be infected with other pathogenic bacteria which may become dangerous in producing secondary and perhaps fatal lesions in animals already diseased. This applies more directly to the swill food which is used by many in the vicinity of large cities and which is composed of such miscellaneous material partly in a condition of fermentation and decomposition that the presence of disease germs may be considered probable at any time. In outbreak IX, in which swill food was mainly used, the bacilli of malignant œdema were obtained from some cases and very likely added to the fatality of the outbreak.

3. THE RELATION OF HOG CHOLERA TO SWINE PLAGUE.*

Throughout this report frequent reference has been made to hog cholera because many outbreaks studied during the past five or six years were mixtures of both diseases, and it is therefore difficult to separate them in order to estimate correctly the damage done by each. We have encountered a small number of outbreaks, of which some were due to hog cholera, others to swine plague, but the majority were the result of a mixed infection.

The outbreaks of hog cholera not complicated with swine plague were generally of a virulent type. When both diseases showed themselves, neither was, as a rule, of any great virulence. To explain the frequent intermingling of these diseases we must refer to the bacteriological results of the past few years. Besides the virulent varieties of hog-cholera bacilli, which produce a characteristic fatal disease in rabbits after subcutaneous inoculation of exceedingly minute doses, we have encountered about half a dozen varieties whose virulence was much diminished. The diminishing pathogenic power is shown by the absence of a fatal disease after subcutaneous inoculation of rabbits, and even small quantities of culture liquid injected into the circulation may produce only a mild disease. In one† there seemed to be no virulence left, and it becomes questionable whether such bacilli can be regarded as hog-cholera bacilli at all. The attenuated bacilli have likewise very little or no effect on swine.

Parallel to this diminishing scale of virulence of hog-cholera bacilli, we have a similar scale among swine-plague bacteria repeatedly set forth in the preceding pages. On the one hand, some varieties will destroy rabbits within 16 hours after inoculation of the minutest trace of culture material into the skin; on the other there are varieties which barely destroy rabbits after large doses have been injected directly

---

* See also pp. 102-103.   † Bacillus γ, outbreak IX, p. 78.

into the circulating blood. All of these varieties have been obtained from the internal organs of diseased swine, and hence even the very attenuated ones may have had some share in the disease.

If we picture to ourselves a wide distribution of these several varieties of hog-cholera and swine-plague bacteria in the bodies of diseased and of partly recovered swine, and, in case of swine-plague varieties, in the air passages of healthy animals, it is not difficult to understand why there are so many mixed outbreaks. The practice already alluded to, of purchasing pigs from many herds and localities and bringing them together to be fattened as one herd, is the most successful method of bringing various grades of pathogenic bacteria together and of producing a mingling of two diseases. These mixed outbreaks may develop in other ways also. The disease may begin as hog cholera and become subsequently complicated with swine plague or the reverse may be true ; the disease may begin as swine plague, and become complicated with hog cholera. In either case the most virulent variety will probably start the disease, and any attenuated hog-cholera or swine-plague bacteria, which are latent in some of the animals of the herd, or have not yet been killed out of the soil, and the surroundings from a former outbreak may start into activity and thus produce a more fatal mixed disease. It is evident that such secondary attacks of attenuated bacteria would not take place if the animals had not been weakened by the primary disease. This may be the only way in which the great majority of the swine-plague bacteria in the air passages of healthy animals can exert any pathogenic effect whatever. It is likewise difficult to understand how attenuated hog-cholera bacilli can act without assistance from swine plague. These statements may be illustrated by referring to the investigations. Thus in outbreak IV the disease was evidently swine plague at first, and complicated with hog-cholera later. For the hog-cholera bacilli were only observed in the later cases. It should likewise be borne in mind that in swine plague some cases are usually of a more chronic type. The disease lasts some time and is associated with caseous changes in the lungs. Any hog-cholera bacilli have thus abundant opportunity to enter the weakened organism and appear after death in cultures from the internal organs. For the same reason hog-cholera outbreaks characterized by very feeble pathogenic activity of the hog-cholera bacilli, and hence of a more prolonged duration and chronic character, are usually complicated with swine plague, because the latter, even though of a feeble activity, has been able to invade the weakened organism and has had time to do so. In virulent outbreaks of either disease death may ensue so rapidly that no invasion of the other disease takes place. These statements presuppose, of course, that both kinds of bacteria exist in the surroundings of the herd.

The appearance of mixed outbreaks due to bacteria brought by different herds is suggested by outbreak VII. Even after a very thorough examination of Nos. 1 and 2 no hog-cholera bacilli could be found. In the subsequent cases in which they were present, they were readily detected in the various organs examined. Moreover there was quite a difference observable in the lesions of the various cases corresponding more or less closely to the nature of the bacteria found. The supposition already presented in regard to this outbreak was that the hog-cholera bacilli were either present in the locality, into which the pigs were brought or were carried by some one or more pigs in the herd.

Hog cholera and swine plague thus mutually assist one another to produce those feebly infectious, chronic diseases which are common at all seasons of the year, and it is not unreasonable to suppose that from such apparently insignificant diseases large epizoöties are developed, which by a gradual return of virulence in the bacteria, and under slight provocations of forced feeding, cold, or other debilitating influences on the part of their victims, burst forth at certain seasons of the year with unexpected violence.*

This condition of things refers more particularly to localities where swine-raising has been conducted on a large scale, where these diseases never actually die out, and where consequently the specific bacteria are always on hand. In those regions which are being invaded by these plagues anew the latter may smolder for several years by reason of the introduction of attenuated varieties before they break out as genuine epizoöties.

There are no facts at hand to indicate any difference in the distribution of these two plagues. The localities where either one or both plagues have been determined by bacteriological investigations may be tabulated as follows:

Locality.	Character of plague.	
By the Bureau of Animal Industry :		
District of Columbia, numerous outbreaks 1885-'90	Hog cholera.	Swine plague.
Maryland, various outbreaks 1885-'90	.. do	Do.
Virginia, various outbreaks 1885-'90	.. do	Do.
Nebraska, 1886	..do	
Illinois (Geneseo), July, 1886		Swine plague.
Illinois (Sodorus), September, 1886	Hog cholera.	Do.
Iowa, December, 1886		Do.
Iowa (Mason City), November, 1888	*(?)	Do.
New Jersey (Johnsonburgh), October, 1887	†(?)	
New Jersey (Pleasantville), July, 1890	(?)	Swine plague.
Missouri (Chillicothe), 1890-'91	Hog cholera.	Do.
Nebraska (1886-'88), by Billings	.. do ..	(?)
Maryland (Baltimore), by Welch and Clement	.. do	Swine plague.
South Carolina, by Bolton	.. do	
Illinois, by Burrill		Swine plague.
Massachusetts (near Boston), by J. A. Jeffries		Do.

* The investigation in Iowa did not bring to light any hog cholera bacilli, though the lesions suggest the presence of attenuated forms not accessible by the usual methods.
† In this small outbreak bacilli closely resembling those of hog cholera were found in the spleen. Their virulence, however, was very feeble. Subcutaneous inoculation had no effect on rabbits.

* A curious instance of this recrudescence of apparently enfeebled virus is outbreak IV referred to above. During February and March swine plague from a neighboring farm became mingled with hog cholera in an infected pen on the Station in which this disease had apparently lost all virulence. The cases of mixed disease were of but slight severity so far as hog cholera lesions were concerned. Gradually towards spring, animals placed in this pen died rapidly of hemorrhagic hog cholera. There was no evidence of the introduction of another bacillus, nor were any changes detected as regards the bacteria in cultures or in the inoculated rabbits.

1. THE RELATION OF SWINE PLAGUE TO DISEASES OF OTHER DOMESTICATED ANIMALS.

The question whether the different species of domesticated animals on a farm may take from or transmit to swine the disease which we have been considering is of very great importance in view of the changing conditions of live-stock interests which are going on in different directions in various parts of our country.

The problem may be stated as follows: Has the bringing together of different species of animals for the purposes of feeding, etc., on the same ground, a tendency to increase disease in one or the other species? Will swine take swine plague from cattle and will they transmit it to sheep and horses, for example, or is the reverse ever observed?

Investigations and observations during the past 15 years lend some color to such possibilities, and it becomes necessary at least to call attention to those engaged in raising and keeping farm animals to what has been determined in this direction, and to arouse their interest in the investigation of outbreaks of swine plague, especially as regards the immediate causes.

In the summer of 1878 there appeared in three royal game preserves, in the vicinity of Munich in Bavaria, a very fatal epizootic among the wild boars and deer, of which 234 boars and 153 deer perished.* It was also noticed that even after the plague in the parks had apparently died out, disease among cattle in the neighborhood appeared, and this, according to the observations of veterinarians, was identical with the disease observed among the game in the parks.

The disease was very acute, lasting from 12 to 36 hours in the majority of cases. In those in which pneumonia and pleuritis were present the disease may have lasted 5 or 6 days. The chief lesions among the latter were croupous pneumonia, pleuritis, pericarditis, and mediastinitis. In cattle the disease appeared in two forms. In one a swelling was observed on the head, the face, the neck, or in the tongue, which assumed enormous proportions in 6 to 12 hours and led to suffocation. The swelling was due to serous or serous and hemorrhagic infiltration. In the other form, in addition to the pneumonia, pleuritis, and pericarditis observed in the game, there was always present a severe hemorrhagic inflammation of the small intestine. Bollinger called these forms exanthematic and pectoral, respectively. At this time bacteriological methods were still undeveloped, and nothing is known of the nature of the bacteria causing this outbreak save the fact that they were not anthrax bacilli. A number of inoculations were made upon various animals, which testify to the extreme virulence of the specific bacteria.

Rabbits died 6 to 8 hours, sheep and goats 30 to 36 hours after inoculation. Two old horses died after subcutaneous inoculation with blood from cattle in a very short time. A young steer, 1½ years old, was fed

* Bollinger. Über eine neue Wild- und Rinderseuche. München, 1878.

with a thimbleful of the intestinal contents of a calf which had succumbed to an enormous swelling. The steer died in 54 hours with
pneumonia and pleuritis. A pig inoculated subcutaneously over the
left shoulder with a few drops of blood died in 22 hours. Besides an
extensive erysipelatous swelling starting from the point of inoculation
there was beginning fibrinous pleuritis.

The disease reappeared in the following years, either sporadically or
in restricted outbreaks. In 1879 and 1880 it was observed among
domesticated animals alone ; in 1881 among the animals in the game
preserves. In 1885 Kitt* published some investigations which were
destined to throw more light upon this new plague. With blood from
an outbreak among cattle resembling the epizoötic described by Bollinger, Kitt made some inoculations upon small animals. Of mice, rabbits, guinea-pigs, and one pigeon inoculated, the mice and rabbits died
within 24 to 36 hours, the pigeon in 36 hours. The guinea-pigs were
not affected. Lesions were in general absent. The blood contained
large numbers of bacteria. Subsequently the spleens of an ox, a young
pig (of which eight had died), and a horse which had succumbed in the
same locality, showed on microscopic examination the same bacteria,
whose virulence tested on rabbits was likewise the same. In a cow inoculated subcutaneously over the left shoulder an extensive inflammatory œdema of the inoculated shoulder appeared, which extended over
the entire left limb. The swelling later became converted into an
abscess, but the animal did not die.

Of special interest is the subcutaneous inoculation of a pig with a
minimum quantity of mouse's blood. From the place of inoculation on
the right thigh a bluish discoloration of the skin spread over the whole
body in spots and patches, while there was considerable swelling at the
place of inoculation. The pig was dead in 24 hours. The autopsy revealed, in addition to the lesions mentioned, exudative pleuritis and
peritonitis, congestion of the mucous membrane of the upper air passages and of the stomach. A goat inoculated subcutaneously in the
same manner was afflicted with extensive local inflammatory œdema and
died within 2 days. A horse inoculated subcutaneously in the neck
with a suspension from an agar culture, derived originally from the
mouse, died within 1½ days with extensive local reaction, fluid blood,
ecchymoses on heart, pleuritis and pericarditis, and beginning inflammation of the mucosa of the stomach.

The bacteria found by Kitt have a marked resemblance to swine-
plague bacteria, and their pathogenic effect on pigs and smaller animals
is identical with that of very virulent swine-plague bacteria. Kitt
states that they grew upon boiled potato as grayish-yellow colonies,
whereas swine-plague bacteria do not produce any visible growth. An

* Ueber eine experimentelle, der Rinderseuche (Bollinger) ähnliche Infectionskrank-
heit. Sitzungsberichte der Gesellschaft f. Morphologie und Physiologie in München,
I, 1885, S. 140-168.

examination of the text leads us, however, to believe that he inoculated potatoes directly with blood. Those who have made many cultures of these bacteria have undoubtedly realized how very richly cultures have grown to which a little blood was transferred from the animal under examination. Hence the potato growths were likely due to the presence of blood. The further statement that, after an examination of Kitt's cultures, Schütz considered these bacteria different from those of *Schweineseuche* (swine plague) because the latter did not kill pigeons is worth nothing, because the difference is simply a matter of virulence. The great difference among swine-plague bacteria themselves as regards this very point we have repeatedly pointed out.

A disease probably identical with the foregoing was described by Oreste and Armanni,* as occurring among herds of young buffaloes in Italy. The disease appears very suddenly, and the animals attacked may die in from 12 to 24 hours. The symptoms are high temperature, rapid and feeble pulse, discharge of mucus from nose and mouth, associated with a local swelling on the head and face which leads to suffocation. The lesions observed after death are few and inconstant, a hemorrhagic inflammation of the small intestine being frequently observed. The specific bacteria seem to be identical morphologically with swine-plague bacteria. The disease can be reproduced in young buffaloes by inoculation of cultures. It was similarly produced in a colt, a cow, a sheep, and in mice, rabbits, guinea-pigs, pigeons, and fowls. Death ensued in all animals in from 1 to 3 days. Of two young pigs inoculated one died, the other survived.

In France, Galtier† has found pneumo-enteritis of swine associated with a similiar disease in sheep which came in contact with them. While there is much in favor of his assumption that the infection passed from the swine to the sheep, the description of the specific bacteria and of the methods of inoculation are not sufficiently complete and thorough to bring conviction as to the transmission, or furnish any definite information concerning the nature of the bacteria found. A few suggestions thrown out here and there are sufficient, I think, to permit us to exclude hog-cholera bacilli and regard them as belonging to the group under consideration.

These various investigations are of great importance in showing that some infectious diseases may either attack several species of domesticated animals at the same time, or be inoculable from one species to another. What is of special significance in the first two investigations is the extreme virulence of the bacteria. The same may be said of the Italian buffalo disease.

There is another class of infectious diseases, due to bacteria of the

* Atti del R. Istituto d' incoraggiamento alle scienze naturali, etc., 1887. For a brief account, see also Journal de Médecine Vétérinaire, 1887, p. 585, and Baumgarten's Jahresbericht for 1887, S. 124.

†Journal de Méd. Vét., 1889, *passim,*

same group, which produce specific diseases among certain species of domesticated animals, but which diseases are not known to be communicable to other species. Among these are fowl cholera, rabbit septicæmia, and a peculiar form of pleuro-pneumonia in cattle, which Poels has called " septic pleuro-pneumonia." During the past 3 or 4 years, the writer has examined in the laboratory of the Bureau of Animal Industry a small number of lungs from cattle affected with pneumonia from which bacteria practically identical with swine-plague bacteria were isolated. A description and discussion of these forms of pneumonia in cattle will be reserved for a future report.

There is thus a wide distribution of diseases among domesticated animals due to a group of bacteria closely resembling and probably identical with swine-plague bacteria. Some diseases attack several species at the same time; others are, so far as we know, restricted to one species. We have also seen that there is a wide distribution of attenuated varieties among the same domesticated animals in the healthy state, inhabiting, so far as our investigations have gone, the upper air passages. Some observers are inclined to regard these different bacteria as practically the same. Hüppe has proposed the name *septicæmia hæmorrhagica* for all the forms of disease caused by them. Other observers hesitate to accept at present this unifying explanation. For practical purposes the following explanation, based on quite extended study of this group of bacteria, may serve as a provisional guide in the prevention of disease.

The real test of the power of any bacteria to produce disease is virulence. The greater the virulence the more liable will be the disease to spread from one species to another. This is strikingly illustrated by the *Wildseuche* of Bollinger. The relative virulence can be accurately determined only by careful series of inoculations upon small and large experimental animals, performed in precisely the same way in each case with pure cultures of the bacteria. Again, the power of a given disease to pass from one species to another frequently remains unnoticed, partly because the opportunity for such transmission is rarely given. Animals of different species, such as swine, cattle, and sheep, are rarely raised and kept in the same inclosures, because the nature of food required for each, and other conditions lead to specialization in stock-raising and tend to restrict each species to its own pasture ground.

It is not unreasonable to suppose that bacteria living in the air passages of one species, and harmless to it to a certain degree, may prove to be disease germs with reference to another species. Thus the attenuated bacteria living in the air passages of healthy cats, dogs, pigs, and cattle, are all fatal to rabbits. In general, the larger and more powerful the animals the less effect disease germs have upon them. It is therefore probable that some of the outbreaks of swine disease in the Western States may be due to the cattle with which the swine are herded for feeding purposes. The bacteria in cattle, harmless to them, or perhaps causing only mild disease and rarely observed, may prove the starting point of disease for swine.

While we have no positive demonstration of these statements, it is desirable that those engaged in stock-raising should have their attention called to the possibilities embodied therein.

5. ON MEASURES TO BE TAKEN IN THE PREVENTION OF SWINE PLAGUE.

In regard to the general measures to be taken and the rules to be observed in the prevention of hog cholera and swine plague, we refer the reader to the report of the Secretary of Agriculture for 1888, page 156, or the report of the Bureau of Animal Industry for 1887–'88, page 148, or the Special Report on Hog Cholera, 1889, page 123. The rules and directions there formulated are adapted as well to swine plague, for the bacteria of the latter disease are even more easily destroyed by various agencies than are hog-cholera bacilli. In the following pages only the most important points are touched upon.

The things with which healthy swine should not come in contact are, in the order of their importance, first of all, diseased herds and animals, strange swine the history of which is not known, offal from establishments using carcasses of swine, recently infected ground, railroads carrying swine, and polluted streams. Soil and water may be infected by living and dead swine or any offal from them.

When the disease has actually appeared in a herd the question generally arises whether it is worth while to make any attempts to save a portion of the herd or to leave them to their fate. As a rule it may be stated that it is best to slaughter both healthy and diseased at once and give the surroundings sufficient time to rid themselves of the infection before fresh animals are brought into them. If this be not desirable we should recommend the following measures to be rigorously carried out:

(*a*) Removal of still healthy animals to uninfected grounds or pens as quickly as possible.

(*b*) Destruction of all diseased animals.

(*c*) Careful burial or burning of carcasses.

(*d*) Repeated thorough disinfection of the infected premises.

(*e*) Great cleanliness both as to surroundings and as regards the food.

If the animals have been removed to uninfected grounds, careful watching is necessary to remove therefrom at once all swine which show signs of disease.

Among the various disinfectants which can be recommended are the following:

1. Slaked lime, in the proportion of about 5 per cent (one-half pound of lime to a gallon of water).

2. Equal volumes of crude carbolic acid and ordinary sulphuric acid mixed together and added to water in the proportion of 2 ounces to a gallon of water ($1\frac{1}{2}$ volume per cent).

3. Sulphuric acid added to water in the proportion of 1 ounce to a gallon.

4. Boiling water.

5. Corrosive sublimate (mercuric chloride) in the proportion of 1 drachm to a gallon of water (1 to 1,000).

Solution No. 2 is said to be more active if, while the sulphuric acid is being added to the crude carbolic acid, the vessel containing the latter is placed in cold water to prevent undue heating of the liquid.

It should be borne in mind that sulphuric acid and corrosive sublimate attack metals, and that the solutions are best made in wooden pails, etc. Corrosive sublimate is also highly poisonous, and the solution should not be made stronger than indicated. The lime is, on the whole, the best and cheapest, but it may not be desirable to use it everywhere; hence, one of the others may be substituted. Each of the solutions recommended is more than strong enough to kill both hog-cholera and swine-plague bacteria and they need not be increased in strength.

When swine have become infected while running over tracts of ground, disinfection of such tracts may be regarded as practically impossible. If, however, they have been brought up in pens or in small inclosures, disinfection should be thoroughly carried out. The woodwork of pens may be disinfected by exposing all portions, cracks and corners, to the action of any of the solutions mentioned. These may be applied with a broom or any other household article which insures uniform wetting. Whitewash is useful for woodwork of fences, etc., when there is no objection to its appearance. Its action is only exerted at the time of application and after it has dried it will not destroy bacteria subsequently adhering to it. It must, therefore, be applied fresh every time disinfection is needed. For large farms some kind of spraying apparatus would be of great service in insuring uniform distribution of the disinfectant. In the selection care must be exercised, however, owing to the corrosive action of some of the solutions. The disinfection of the surface of the soil over small areas is perhaps best accomplished by the slaked lime or the crude carbolic-acid solution. It should be remembered that both preparations may be irritating to the feet of animals immediately after they have been applied. The feeding troughs should receive special attention, and after the application of the disinfectant this should be washed away with water, preferably hot or boiling.

The directions thus far given apply mainly to the prevention of disease. When animals have been actually attacked, can anything be done for them? It has already been stated that treatment of communicable diseases is not a desirable thing, but even if it were the deaths follow each other so rapidly in many outbreaks that there is no time for the application of remedies. If, however, an effort to treat them is to be made, it is desirable to avoid the various specifics and remedies of unknown composition, some of which, thoroughly tested at the Bureau Station by Dr. F. L. Kilborne, were of no avail in checking the disease.

The sick animals should be isolated one from another, as far as possible confined in small inclosures, kept quiet, and fed with moderate quantities of food, preferably with milk, if this is to be obtained. If the swine are being fattened when the disease appears, this process should be stopped at once and a light diet substituted. The tendency towards the localization of disease in the large intestine, in both swine plague and hog cholera, seems to be due, at least in part, to the constipated habits of the pig, which permit the pathogenic bacteria to remain long enough in the intestine to act injuriously upon the mucous membrane. Constipation is not easily overcome, as the trials with various cathartics * have demonstrated, and it is highly important when the disease has appeared to feed a greater variety in small quantity or to follow the recommendation of giving the digestive organs a complete rest by feeding milk. The boiling of food may be desirable, inasmuch as it destroys any disease-producing bacteria which may be present, and makes digestion easier. An experiment carried out at the Bureau Station with boiled food did not show any more favorable results, however, than with unboiled food ordinarily given, so that we can simply suggest it for further trial.

Even if treatment should succeed after much trouble and expense to save some few swine, it may not be profitable, owing to the injury inflicted on the various organs during the disease. The lungs are, as a rule, seriously affected. They may become adherent to the walls of the thorax, and the pericardium may become firmly attached to the heart and impede its action. These permanent injuries, which no kind of treatment yet suggested can avoid, exercise an injurious influence on the proper development of the animal affected and make its raising of questionable advantage. It has already been stated that such recovered animals may for a time at least be dangerous as carriers of the disease germs to other swine.

The only encouraging line of action, therefore, lies in the prevention of disease by the observance of suitable precautionary measures and in that general practice of hygienic laws which thus far has been the only means of checking the rapid spread of epidemics in the human family. The method first suggested by Pasteur of inoculating animals with attenuated cultures, to make them resist any and every attack of a given infectious disease, is, theoretically considered, the simplest means of prevention. Practically, however, there are two objections which are growing in importance year by year, as our knowledge of infectious diseases is becoming broader and deeper. The method of Pasteur may distribute the specific bacteria far and wide and become a source of future evil, since we do not know but that the attenuated bacteria may in some way regain their former virulence. The other objection rests on the fact that diseases differ so much one from the other that the method seems to insure success in only a few diseases.

---

* Special Report on Hog Cholera, 1889, p. 135.

As regards swine plague, the experiments which have thus far been carried out indicate that this disease may prove amenable to preventive inoculation. We have been able, by the injection of both living cultures and those sterilized at a low temperature (58° C.), to make the most susceptible animals, rabbits, insusceptible to the most virulent swine-plague bacteria. By two subcutaneous injections of cultures of swine-plague bacteria swine have been made insusceptible to doses injected into the circulation, which proved fatal to "control" pigs within 24 hours. In the preliminary experiments upon rabbits, designed to produce immunity, several methods were employed.*

1. Minute but gradually increasing quantities of culture liquid of very attenuated swine-plague bacteria were injected at different intervals into the ear vein of rabbits. Only a very small proportion of these survived the test inoculation with very virulent swine-plague bacteria.

2. Sterilized bouillon cultures were injected into the abdomen and into the circulation of rabbits. This method also produced immunity and partial resistance, but in only a comparatively few animals.

3. The preceding method was modified in the following manner: Swine-plague bacteria from outbreak IX were allowed to produce for 2 days a rich growth upon agar. This growth was scraped off and a very turbid suspension in bouillon prepared and sterilized at 58° C. With this sterilized suspension injections were made into the abdomen of rabbits as follows:

Rabbit No.	May 4.	May 8.	May 14.	May 22.	Total.	Remarks.
	cc.	cc.	cc.	cc.	cc.	
35	1.5	1	2	3	7.5	Nos. 35, 36, and 37 inoculated with virulent swine plague May 26. Check dies over night. All three survive with considerable local reaction.
36	2	.5	2	3	7.5	
37	1	1.5	2	3	7.5	
38	.5	2	2	....	4.5	Inoculated with virulent swine plague May 19, dies in 6 days with severe local reaction, pleuritis, and pericarditis. The check dies in 16 to 20 hours.

These results show very decisively the protective effect of the sterilized growth of swine-plague bacteria. Additional experiments have not yet been made.

In conjunction with Dr. Kilborne, the protective effect of swine-plague cultures was tested upon swine in the following experiment: Seven pigs belonging to the same lot and about 4 months old were chosen, three of which were set aside as "control" animals or checks. The remaining four received February 28, 1891, a subcutaneous injection of 6 cubic centimetres of peptone-bouillon culture of virulent swine-plague bacteria (outbreak IX), one-half into each thigh. As a result one died.† The remaining three were reinoculated in the same way March 14,

---

* These experiments were carried out in conjunction with Dr. V. A. Moore.

† See p. 74 for autopsy notes.

receiving on this date 10 cubic centimetres of culture liquid. April 3, these, together with the three control animals or checks, received the final test inoculation: 2 cubic centimetres of peptone-bouillon culture of the same bacteria were injected into a vein of the leg of each animal. Two of the control animals died within 24 hours, the third in 36 hours. None of the three vaccinated animals became ill. No symptoms of disease or lesions appeared subsequently.

These experiments simply demonstrate the fact that swine may be protected from fatal doses by subcutaneous injection. Whether this process would be successful in natural outbreaks can not be inferred from this test. The method is open to the objection above mentioned, i. e., it is liable to distribute the specific bacteria wherever vaccination is practiced. Since the more desirable one of injecting the products of bacterial growth is now being tested there is no need of any further discussion of this subject at the present time.

CONCLUSIONS.

1. There are two independent infectious diseases of swine—swine plague and hog cholera—each caused by an easily recognizable, specific disease germ.

2. Swine plague (in those outbreaks which have come to our notice) is limited chiefly to the lungs in its destructive effect. The intestines may be and frequently are involved in the disease process. Hence it is an infectious pneumo-enteritis rather than an infectious pneumonia.

3. There is considerable variation in the virulence or disease-producing power of swine-plague bacteria from different outbreaks. The greater the virulence, other things being equal, the severer and more extensive the epizoötic.

4. The bacteria of Schweineseuche (German disease of swine) are identical with those of swine plague.

5. In the upper air passages of a certain percentage of healthy swine, cattle, dogs, and cats, bacteria exist which belong to the species of swine-plague bacteria, and which as a rule possess a relatively feeble virulence. While it is probable that such bacteria may produce disease it may be regarded as pretty certain that it is largely aided by secondary causes producing unthriftiness, and is merely sporadic and not communicable.

6. In many epizoötics of swine disease both hog-cholera and swine-plague bacteria as well as the respective lesions of these bacteria coexist. Such mixed diseases are due to the frequent presence of both bacteria in the surroundings of swine, probably a result of frequent introduction. Either disease may be primary according to its relative virulence.

7. It is highly probable that the many attenuated varieties of either disease germ can produce disease only when assisted by the other germ

or by the unsanitary, unphysiological methods of rearing swine by which the latter are reduced in vitality and made more susceptible.

8. It is pretty well established that there are a number of infectious diseases affecting cattle, buffaloes, deer, fowls, and smaller animals, the bacteria of which are closely related, if not identical with, those of swine plague. These plagues appear in various parts of the globe sporadically. (*Wild- und Rinderseuche, barbone bufalino,* fowl cholera, rabbit septicæmia.) Their tendency to spread from one species to another, from cattle to swine, for instance, probably depends both on the degree of virulence of the bacteria as well as the opportunities afforded for such transmission.

9. Swine-plague bacteria are very probably introduced into a herd only in the bodies of animals, since they are speedily destroyed in soil and water by natural agencies. Virulent varieties are perhaps always derived from preëxisting disease. Attenuated varieties may be introduced by healthy animals. Since these may under special conditions give rise to disease, efforts to prevent and suppress infection must take into account the physical condition of the exposed animals.

APPENDIX.

THE PRESENCE OF SEPTIC BACTERIA, PROBABLY IDENTICAL WITH THOSE OF SWINE PLAGUE, IN THE UPPER AIR PASSAGES OF DOMESTICATED ANIMALS OTHER THAN SWINE.

By VERANUS A. MOORE, B. S., M. D., *Assistant.*

The examination of the secretions of the mucous membrane of the upper air passages in domesticated animals other than swine was begun under the direction of Dr. Theobald Smith for the purpose of determining whether or not the swine-plague germ, or a germ closely related to it, is normally present in these animals. The results obtained from the limited number of examinations that have been made from the various animals are of so much value in throwing light upon the natural habitat of this group of microörganisms that a preliminary report of these experiments seemed desirable at this time.

The methods that have been employed in these investigations are the same as those used by Dr. Smith in the examination of mucus from the respiratory tract of healthy pigs, and which are described on p. 110 of this report. The inoculations of rabbits with the mucus from the various animals were made in part conjointly with Dr. F. L. Kilborne, in part by him alone, and in a few cases I alone am responsible for these operations.

1. *Inoculations from cattle.*—Rabbits have been inoculated with the mucus taken from the larynx or amygdaloid cavities of seven healthy cattle. Four of these were steers, two of which were Western animals that had been shipped to Washington for beef. The four steers were killed in a slaughter-house near the Experiment Station by cutting the blood vessels in the neck. Care was taken to keep the mouth free from blood. The other three were heifers that were killed for various purposes in a similar manner at the Experiment Station. In each case the mucus was collected immediately after the death of the animal and inoculated subcutaneously into rabbits.

The rabbits that were inoculated with the mucus from the four steers and one of the heifers died in from 3 to 6 days. The lesions found in these rabbits were similar to those produced by the attenuated swine-

151

plague germ. The local infiltration contained several forms of bacteria. In the peritoneal exudate and in the spleen and liver bacteria were found that resembled the swine-plague germ both in stained cover-glass preparations from the tissues and in cultures.

The rabbits inoculated from the two remaining animals showed no signs of disease.

The pathogenic effect of the bacteria obtained from the first two cattle was determined by inoculating rabbits with pure cultures. The subcutaneous injection produced extensive purulent infiltration at the point of inoculation and exudative peritonitis, destroying the rabbit in 6 days. An intravenous inoculation of the same quantity of a similar culture from the second case proved fatal in 24 hours. The blood, liver, and spleen contained innumerable bacteria which could not be distinguished from the swine-plague germ. No inoculations were made with pure cultures from the remaining three animals. It is sufficient to say that the lesions produced in all of the rabbits inoculated from the five cases were the same. The following tables will explain the results of these inoculations:

Inoculation of rabbits with mucus from the upper air passages of cattle.

Animal No.	Mucus from	Rabbit inoculated, No.	Date of inoculation.	Rabbit died in—	Remarks.
			1890.	*Days.*	
1	Amygdaloid cavities.	1	Feb. 27	3	Local reaction; peritonitis.
		2	Feb. 27	4	Do.
2		3	Feb. 27	4	Local reaction; peritonitis; pleuritis.
	Larynx ...	4	Feb. 27	4	Local reaction; peritonitis; pleuritis and pericarditis.
3	Amygdaloid cavities.	5	Mar. 13	3	Local reaction; peritonitis.
4	... do	6	Apr. 1	4	Local reaction; peritonitis; beginning pleuritis.
5	... do	7	Mar. 20	....	Rabbit remained well.
6	Larynx ...	8	Oct. 2	4	Local reaction; peritonitis.
		9	Oct. 2	6	Do.
			1891.		
7	Amygdaloid cavities.	10	Jan. 5	....	Rabbit remained well.
		11	Jan. 5	....	Do.

Inoculations with pure cultures obtained from above rabbits.

Culture from rabbit No.	Method of inoculation and date.	Rabbit died in—	Remarks.
		Days.	
1	March 1, 1890, one-eighth cubic centimetre bouillon* culture subcutaneously.	6	Local reaction; peritonitis.
3	March 4, 1890, one-eighth cubic centimetre bouillon culture into ear vein.	1	Septicæmia.

* All the bouillon used in these investigations contained one-fourth per cent. of peptone.

2. *Inoculations from cats.*—Rabbits were inoculated with the secretions of the mucous membrane of trachea, larynx, or pharynx of seven

healthy cats. The cats were raised in and about Washington, but not on the Experiment Station. They were killed either by a shot through the heart or with chloroform, and the mucus was removed with every precaution immediately after death.

The rabbits that were inoculated from cat No. 6 remained well. All of the others died. The result of these inoculations is of particular interest, as the rabbits died in from 1 to 7 days, and presented lesions similar to those produced by the swine-plague germ in its most virulent as well as its attenuated forms. From the various organs of all the rabbits bacteria were found which could not be distinguished from each other or from the swine-plague germ.

The virulence of pure cultures obtained from the rabbits inoculated from cats 1, 2, and 3 was tested by both intravenous and subcutaneous inoculations on fresh rabbits. These proved fatal in from 18 to 48 hours. The blood and other organs contained innumerable bacteria. The cultural characters of these germs will be mentioned in another place. The accompanying tables give a summary of these inoculations:

Inoculation of rabbits with mucus from the upper air passages of cats.

Cat No.	How killed.	Mucus from—	Rabbit inoculated, No.	Date of inoculation.	Rabbit died in—	Remarks.
				1890.	*Days.*	
1	Shot through heart.	Larynx .	12	Apr. 22	15	Local reaction; peritonitis
			13	Apr. 22	7	Local reaction; peritonitis pleuritis, pericarditis.
2	....do ....do		14	Apr. 22	2	Local reaction; peritonitis.
			15	Apr. 22	3	Do.
3	.. do .... Pharynx		16	June 11	1	Septicaemia.
4	....do ....do		17	June 11	1	Do.
5	....do .... Larynx ..		18	Dec. 26	1½	Slight local reaction, beginning peritonitis.
				1891.		
6	Chloroformed Trachea .		19	Jan. 6	....	Rabbit remained well.
			20	Jan. 6	....	Do.
7	....do ....do		21	Jan. 21	3	Local reaction; beginning pleuritis.
			22	Jan. 24	5	Local reaction; pleuritis.

Inoculations with pure cultures obtained from above rabbits.

Culture from rabbit No.	Method of inoculation and date.	Rabbit died in—	Remarks.
		Hours.	
12	Apr.29,1890, one-eighth cubic centimetre bouillon culture into ear vein.	20	Septicaemia.
	May 2, 1-90, loop agar culture subcutaneously in ear.	60	Beginning peritonitis.
14	May 10,1890, one-eighth cubic centimetre bouillon culture into ear vein.	18	Septicaemia.
15	May 10,1890, one-eighth cubic centimetre bouillon culture subcutaneously.	36	Slight local reaction; septicaemia.
16	.. do	18	Septicaemia.

3. *Inoculations from dogs.*—Rabbits were inoculated with the mucus taken from the larynx or upper pharynx of six healthy dogs. These were also procured in Washington City and its suburbs. They were killed by a shot through the heart. The mucus was removed immediately after death and at once inoculated.

The rabbits inoculated from dogs Nos. 2 and 3 died in about 36 hours. Innumerable bacteria were found in the various organs that could not be distinguished from those obtained from cattle and cats or from the swine-plague germ. The rabbits inoculated from the other four dogs remained well. Both subcutaneous and intravenous inoculations of fresh rabbits with cultures of these bacteria proved that they were as virulent as those from several of the cats. The results of these inoculations are summarized in the appended tables:

Inoculation of rabbits with mucus from the upper air passages of dogs.

Dog No.	Mucus from—	Rabbit inoculated, No.	Date of inoculation.	Rabbit died in—	Remarks.
			1890.	Hours.	
1	Larynx ... {	23	Apr. 25	....	Rabbit remained well.
		24	Apr. 25	....	Do.
2	Pharynx..	25	Apr. 28	36	Local reaction, beginning pleuritis.
3	Larynx ...	26	May 9	36	Local reaction, peritonitis.
4	....do	27	May 26	....	Rabbit remained well.
5	...do	28	May 31	....	Do.
6	....do	29	Dec. 0	....	Do.

Inoculations with pure cultures obtained from above rabbits.

Culture from rabbit No.—	Method of inoculation and date.	Rabbit died in—	Remarks.
		Hours.	
25	May 2, 1890, loop agar culture subcutaneously in ear.	22	Septicæmia.
26 {	May 20, 1890, one-eighth cubic centimetre bouillon culture into ear vein.	18	Do.
	May 20, 1890, one-eighth cubic centimetre bouillon culture subcutaneously.	18	Do.

4. *Inoculations from other animals.*—Rabbits have been inoculated with the mucus taken from the larynx or trachea of one sheep, one horse, two old fowls, and one rabbit. One of the two rabbits inoculated from the sheep developed a large abscess near the point of inoculation. It was chloroformed after about one month. There were no other lesions. The other rabbits remained well. The inoculations from a single animal are, of course, insufficient to give any general information respecting the species. The annexed table gives all the information necessary with reference to these inoculations.

Inoculations with mucus.

Animal.	How killed.	Mucus from—	Rabbit inoculated, No.	Date of inoculation.	Remarks.
				1890.	
Sheep 1 ...	Cutting jugulars..	Larynx ... {	30	Apr. 7	Local abscess; chloroformed.
			31	Apr. 7	Rabbit remained well.
Horse 1....	Shot	....do .... {	32	Apr. 7	Do.
			33	June 3	Do.
Fowl 1....	Neck broken	Trachea ..	34	June 3	Do.
Fowl 2....	....do	....do	35	June 3	Do.
				1891.	
Rabbit 1 ..	Chloroformed	....do	36	Apr.	Do.

It will be seen from the experiments given that a germ which is not distinguishable from the swine-plague germ was found in the mucus from the upper air passages of 71 per cent. of the cattle, 85 per cent. of the cats, and 33 per cent of the dogs from which inoculations were made. When the rabbit lived for more than 24 hours after its inoculation there was a purulent infiltration of the skin and subcutis at the point of inoculation. The infiltration extended over an area varying in size in proportion to the length of time which the animal lived. In some cases it covered the entire ventral aspect of the body. Occasionally there was in addition a sanguinolent effusion which extended beyond the limits of the infiltration. The local reaction was undoubtedly increased by the presence of other bacteria that were introduced with the mucus. The internal lesions were characterized by an inflammatory condition produced by the localization of the germs on some one or more of the serous membranes, notably the peritoneum, when the rabbit did not die from a rapidly fatal septicæmia.

It is interesting to note that the inoculations made from cattle proved fatal in from 3 to 6 days, and that the resulting lesions in every instance were characteristic of attenuated swine plague. In the rabbit that lived 6 days there was severe peritonitis, while in three that lived only 4 days there were both peritonitis and pleuritis and in one case pericarditis. The rabbit which lived 6 days after inoculation with a pure culture exhibited in addition to the local reaction only peritonitis.

In the inoculations from cats we find a much wider range in the character of the lesions produced. The rabbits inoculated from cats Nos. 3 and 4 were victims of a rapidly fatal septicæmia, the swine-plague bacteria being distributed in enormous numbers throughout the blood and internal organs. The rabbit inoculated from cat No. 5 lived about 12 hours longer and exhibited beginning peritonitis. The localization of these germs on the peritoneum is further illustrated in rabbits Nos. 12, 14, and 15. These animals lived 2 and 3 days and died with exudative peritonitis. In the exudate there were innumerable bacteria, but comparatively few were found in the blood. In rabbit 13 we have a marked example of the distribution of these germs over the entire serous surfaces of the trunk. Both rabbits from cat No. 7 are interesting, as the lesions were confined to the pleura. In these cases the pleuritic exudate contained innumerable swine-plague bacteria while the blood contained only a few. This emphasizes the fact that in cases of well-marked localization there are comparatively few germs in the general circulation at the time of death.

The rabbits that were inoculated from dogs Nos. 2 and 3 lived the same length of time as the rabbit inoculated from cat No. 5. It is interesting to note that the rabbit inoculated from No. 2 exhibited pleuritis and the one from dog No. 3 peritonitis. Here again we have a marked illustration of the variable localization of this group of micro-organisms when their virulence is not sufficient to destroy the rabbit in from 18 to 24 hours. The tendency to localization is well shown in the following summary of the lesions found in the nineteen rabbits that have succumbed to the inoculations from the different animals:

Summary.

Lesions.	No. of rabbits.	Per cent.
Local reaction, peritonitis	10	52.6
Local reaction, pleuritis	3	15.7
Local reaction, peritonitis, and pleuritis	2	10.5
Local reaction, peritonitis, pleuritis, and pericarditis	2	10.5
Septicæmia	2	10.5

The inoculations with pure cultures of the bacteria obtained from the different rabbits, although few in number, are important, as they verify the results obtained from the original inoculations. The subcutaneous inoculation with one-eighth cubic centimeter of a bouillon culture of the germ from cattle resulted in extensive local reaction and peritonitis. The inoculation of a rabbit subcutaneously in the ear with a loop of an agar culture from cat No. 1 is also interesting, as it not only lived nearly as long as the original rabbit, but developed peritonitis. The rabbit inoculated subcutaneously with a culture from cat No. 2 is the only rabbit in these experiments that lived over 24 hours without exhibiting some point of localization of the germs on the serous membranes.

The difference in the virulence of the germs obtained from cattle, cats, and dogs, and the consequent variations in the character of the lesions produced in rabbits, are paralleled in similar inoculations from pigs and with cultures of the swine-plague bacteria obtained from sporadic cases and the different outbreaks of that disease, I have, therefore, not found in the inoculations of rabbits any pathogenic property possessed by any of the septic germs discovered in the upper air passages of healthy animals that will differentiate them from the swine-plague bacteria.

Cultural characters.—From the blood or spleen of each rabbit cultures were made on agar or in bouillon. From these, subcultures were made in the various culture media employed in differentiating bacteria for the purpose of determining, if possible, any cultural differences that might exist between them, or between them and the swine-plague bacteria. As these cultures were obtained at different times and the media used prepared on different dates the occasional slight variations in the character of the growth that were observed between the different bacteria could not be considered as constant differences, as it was found that these bacteria, like those of swine plague, do vary slightly in culture media. In view of this fact two series of comparative cultures have been made on the different media, each medium being prepared from the same material and at the same time.

The comparative cultures were made (1) from cultures of bacteria obtained from a healthy pig, cat, and dog; (2) from cultures of attenuated and of virulent swine-plague bacteria, and (3) from cultures of swine plague bacteria found in a guinea-pig that died of sporadic pneumonia. Unfortunately the germs from the upper air passages of cattle had perished at the time the comparative cultures were made.

(a) *Nutrient agar.*—The growths of the various bacteria upon this substratum were not distinguishable one from the other.

(b) *Alkaline peptonized bouillon.*—The growth in the bouillon cultures made from the blood or spleen of a few of the rabbits consisted of small, grayish flakes. These were at first held in suspension in the liquid, but soon settled, leaving the supernatant culture fluid perfectly clear. This character was not constant, as the clumps of growth gave way to a uniform cloudiness of the culture liquid after a short series of subcultures. The bouillon cultures from all of the other rabbits were uniformly clouded.

In the first series two of the germs grew in clumps. The others imparted a uniform cloudiness to the liquid. After 7 days standing the growth had settled in the bottom of the tube in the two cultures that contained clumps. In the others a thin, grayish, somewhat viscid band composed of bacteria was formed on the sides of the tube at the surface of the liquid. The latter was faintly clouded. The growth of the virulent swine-plague germ seemed less vigorous than that of the others. In about 3 weeks there was a considerable quantity of a grayish, viscid sediment in the bottom of the tubes which upon agitation was forced up, appearing as a somewhat twisted, tenacious cone with its apex at the surface of the liquid.

In the second series the growth in all of the cultures imparted a uniform cloudiness to the liquid. In 48 hours the virulent swine-plague culture was nearly cleared. In 7 days the cultures of the bacteria from the healthy pig and cat and from the guinea-pig's lung were clear. The cultures of the attenuated swine-plague germ and the germ from the healthy dog remained clouded. In every case the grayish band formed on the sides of the tube at the surface of the liquid as in four of the cultures in the first series. The sediment in the bottom of the tubes was small in quantity and friable. The reaction of all of the cultures was decidedly acid after 24 hours; less strongly so after 1 weeks. The difference in the character of the sediment in the two series of cultures was very marked. This same variation has been observed in other bouillon cultures of the same bacteria. It is important to add that the variation in the character of the growth in bouillon cultures of any one of these germs has been found to be as great as that between cultures of the bacteria from different sources.

Although these bacteria change the reaction of an alkaline bouillon to an acid one during their multiplication, they will not grow when inoculated into peptonized beef-broth that has not been neutralized.

(c) *Peptonized bouillon containing 2 per cent. glucose.*—The growth of the various bacteria in fermentation tubes containing this liquid does not vary in the cultures examined.

(d) *Gelatine.*—The growth in this medium is uncertain. The germ from cat No. 2 developed minute grayish colonies in roll cultures. They were, however, too small for diagnostic purposes. One of the swine-plague germs occasionally developed minute colonies. The other bacteria did not grow, although a large number of cultures were made from each.

(e) *Potatoes.*—No growth.

(f) *Milk.*—No appreciable change in the appearance of the milk was produced. Slightly acid in reaction. Cover-glass preparations showed a vigorous multiplication of the bacteria in every culture. In this medium the bacteria appeared as rods longer than under other conditions under which they have been examined.

From both the comparative cultures and the large number of cultures that have been made at other times, I have thus far been unable to detect any cultural character that is sufficiently constant to differentiate the bacteria in question, the one from the other.

PNEUMONIA AND PLEURITIS IN A GUINEA-PIG CAUSED BY BACTERIA CLOSELY RESEMBLING THOSE OF SWINE PLAGUE.

On December 24, 1890, a large adult female guinea-pig was found dead in a pen where several of the supply animals were kept. A careful examination of this animal revealed the following conditions:

Beneath the skin near the left mamma a closed abscess. Swelling and ulceration of the left fore foot. Spleen normal. Liver fatty. No intestinal lesions. In the pleural cavity a considerable quantity of a grayish, viscid exudate lining the costal and pulmonary pleura. The anterior half of both lungs hepatized. Suppurative pericarditis. Cover-glass preparations from the pleural exudate contained a very large number of bacteria not distinguishable from swine-plague bacteria. In an agar tube inoculated with the exudate a pure culture of these bacteria developed.

A rabbit inoculated subcutaneously with a very small quantity of the pleural exudate died in less than 20 hours. Innumerable swine-plague bacteria were found in the liver, blood, and spleen. The polar stain so characteristic of the swine-plague germ was well marked in the cover-glass preparations from all of the tissues. A pure culture of this germ was obtained from the blood.

In order to test still further the virulence of this germ, a second rabbit was inoculated subcutaneously with an equivalent of one-five hundredth cubic centimetre of a bouillon culture made from the blood of the first rabbit. This inoculation proved fatal in about 24 hours. Innumerable swine-plague bacteria were found in the various organs.

The cultures of this germ on the various media could not be distinguished from those of the virulent swine-plague germ.

Dr. Kilborne informs us that prior to the death of this guinea-pig, of which a bacteriological examination was made, others had been found dead from time to time at the Experiment Station without exposure to any disease. A cursory examination had shown that a considerable number of these had died from exudative pleuritis with or without pneumonia.

DISEASE IN A FOWL ASSOCIATED WITH BACTERIA CLOSELY RESEMBLING THE FOREGOING.

On April 21, 1891, two large, well-nourished hens were found dead in a flock of fowls that are kept on the Experiment Station. A few days prior to this a fowl had died, but it was not examined. There had been no evidence of a contagious disease in the flock up to this date, and no deaths occurred subsequent to the ones reported here. A careful examination of these fowls showed that one contained bacteria closely resembling swine plague, and that in the other there was an extensive cronpous exudate throughout the larynx and trachea, the specific cause of which was not determined.

Fowl 1. This fowl was sick 2 days before its death. Heart muscle pale; in right ventricle a mixture of a pale and dark gelatinous clot. Liver sprinkled with a few grayish spots, apparently necrosed tissue. Spleen normal. Kidneys injected with urates and enlarged. Trachea and œsophagus normal. On cover-glass preparations from the liver are minute bodies which appeared to be bacteria. Tubes of agar inoculated with a bit of the blood and liver developed a rich grayish growth not distinguishable from an agar culture of swine-plague bacteria.

April 24 a rabbit was inoculated subcutaneously with a very small quantity of the growth from the blood culture. The rabbit died in 20 hours. Innumerable swine-plague bacteria in the various organs. The polar stain was very marked in stained cover-glass preparations. Pure cultures of the same bacteria were obtained from the blood.

April 28 two fowls were inoculated with an agar culture from the blood of fowl No. 1. The surface growth of the agar culture was diluted with about 1 cubic centimetre of sterilized bouillon. Of this dilution, fowl *a* received 0.5 cubic centimetre subcutaneously over the pectoral muscle, and fowl *b* received 0.5 cubic centimetre into the pectoral muscle.

Fowl *b* died May 4. At point of inoculation a yellowish membranous sequestrum beneath the skin over an area about 2 inches long ; beneath this the pectoral muscle was necrosed to a depth of about one-half inch. The surrounding muscle was sprinkled with punctiform hemorrhages. Heart muscle pale ; considerable serum in pericardial sac, which contains also several small straw-colored coagula. Liver fatty ; somewhat mottled, with grayish and bright red areas. Spleen enlarged ; friable. Kidneys pale ; fatty. The mucous membrane of intestines somewhat injected. Œsophagus and trachea normal. Lungs of a grayish color ; not consolidated. A very few oval bacteria, which did not exhibit the polar stain, were found in the spleen, liver and blood. Cultures from the blood and liver could not be distinguished from cultures of swine-plague bacteria.

Fowl *a* found dead May 11. Fowl much emaciated. At the point of inoculation a sequestrum about 1 inch long lying beneath the skin; subjacent muscle reddened. Heart muscle pale. Liver fatty ; quite firm. Spleen friable. Kidneys dark. The mucous membrane for a distance of about 4 inches below duodenum in the small intestine has a roughened appearance, resembling superficial necrosis. A few oval germs found in liver and blood. An agar culture from the liver showed same characters as cultures from fowl *b*.

From the agar culture of the blood of fowl 1, agar and gelatine plate cultures were made. The agar plates developed colonies not distinguishable from swine-plague colonies. The gelatine plates remained free from growth. Other cultures have been made on the various media, but thus far no difference has been detected between the growth of this germ and that of swine plague.

Fowl No. 2. This fowl was not known to have been sick. Sternum showed evidence of an old injury. Liver fatty. Spleen enlarged. Kidneys normal. Mucous membrane of duodenum generally reddened, also sprinkled with minute bright red dots, probably injected villi. The mucous membrane of the mouth and œsophagus swollen, hyperæmic, and covered with a thin yellowish very friable exudate. The follicles and glands deeply reddened. The larynx and trachea contained a yellowish, croupous exudate in the form of a tube, easily removable in sections from one-fourth to 1 inch in length. The mucous membrane beneath the exudate swollen ; cyanosed. It does not extend into the bronchi. Lungs normal. No bacteria were found in the liver, spleen and blood. Cultures made from these organs remained clear. Agar plate cultures were made from the tracheal exudate. There developed a few chromogenic colonies, and about five colonies of a very slender motile bacillus.

Two rabbits inoculated subcutaneously with pieces of the exudate remained well.

A rabbit inoculated in ear vein with 0.5 cubic centimetre of a bouillon culture of the bacillus obtained from the agar plates, and two mice inoculated subcutaneously with the same culture, remained well.

DESCRIPTION OF PLATES.

PLATE I:

Lungs of a healthy pig inflated, viewed from the left side: *a*, principal lobe: *b*, ventral lobe; *c*, cephalic lobe; *e*, apex of heart. The dotted area bounded by the line *xy* indicates the portion usually involved in disease.

PLATE II:

The same lungs viewed from beneath (ventral, diaphragmatic surface); *a*, principal lobe; *b*, ventral lobe; *c*, median or azygos lobe belonging to the right lung; *e*, apex of heart. The dotted area shows the average extent of the disease.

PLATE III:

Lateral view of right lung of pig No. 407, outbreak IV. (See p. 24). The hepatized regions are almost completely covered with a false membrane.

PLATE IV:

The same lung as seen from the ventral surface. A portion of the diaphragm is fastened to it by means of exudate. The localization of the disease in the anterior (cephalic) and ventral portions is well brought out in these two plates.

PLATE V:

Right lung from case 9, outbreak VII (see p. 38), showing hepatization of portion of cephalic, ventral, and adjacent principal lobe. Minute necrotic masses disseminated through the hepatized tissue. On the left more recent disease with marked interlobular œdema.

PLATE VI:

Left lung of No. 275 (p. 46), showing extensive pneumonia after the injection of culture liquid into the right lung. There is in addition exudative pleuritis and pericarditis.

PLATE VII:

Section of lung passing through bronchus. In the principal lobe around bronchus the lung tissue is completely transformed into firm caseous masses. From outbreak IV.

PLATE VIII:

Fig. 1. Section through one of the lobes of a diseased lung from outbreak IV, illustrative of the caseation so frequently encountered in this outbreak. The irregular patches of a homogeneous, faintly yellowish tint represent the cut surfaces of caseous masses.

Fig. 2. A portion of the mucous membrane of the large intestine (outbreak IV), showing the peculiar isolated masses of exudate found in early cases of this outbreak.

1614——11   161

PLATE IX:

Fig. 1. Collapse of groups of lobules in the principal lobe of a pig's lung. Frequently associated with bronchitis and lung worms.

Fig. 2. Broncho-pneumonia. The cut surface of the lung tissue shows the occluded small air tubes as yellowish spots. The air vesicles or alveoli appear as minute yellowish dots in groups on the surface of the lungs, the color being due to the cell masses filling them up. The exudate plugging the air tubes is sometimes firm enough to be teased out in the form of branching cylinders. This form of lung disease is frequently associated with both hog cholera and swine plague, and may occur independently of them.

PLATE X:

Heart exposed by removing pericardium. The surface of the heart (epicardium) is covered with exudation. The pericardium very much thickened by exudation of similar character. From case 12, outbreak VII.

PLATE XI:

Fig. 1. Cover-glass preparation from spleen of rabbit inoculated with a particle of lung tissue from case 15, outbreak IX. Rabbit dead within 40 hours. Preparation stained in alkaline methylene blue and mounted in Xylol balsam. × 1,000.

Fig. 2. Section from left lung of inoculated pig No. 275, outbreak VII (see p. 46), showing extensive cell infiltration of the alveoli and small air tubes. Tissue hardened in alcohol, stained in alkaline methylene blue. Mounted in balsam. × 140.

Fig. 3. A portion of the contents of an alveolus from the preceding figure highly magnified to show swine-plague bacteria. × 1,100.

Fig. 4. From the liver of No. 454 inoculated subcutaneously with bacteria from outbreak IX (see p. 74). Intralobular capillary containing a mass of swine-plague bacteria. Section prepared and stained as indicated in the description of the preceding plate. × 1,100.

PLATE XII:

Fig. 1. *a*, Surface colonies and deep colonies of swine-plague bacteria (outbreak IX) on an agar plate one week old. The small, round, and lenticular bodies represent the deep colonies, the larger ones the surface colonies; natural size. *b*, A surface colony enlarged 17 diameters, showing reticulated center, and delicate radially striated periphery.

Fig. 2. Two deep colonies from the same plate enlarged 17 diameters.

Fig. 3. Colonies of swine-plague bacteria from outbreak I (1886), on a gelatine plate, 7 days old. × 60.

Fig. 4. Agar tube culture of swine-plague bacteria (from outbreak VIII), about 2 days old. Natural size.

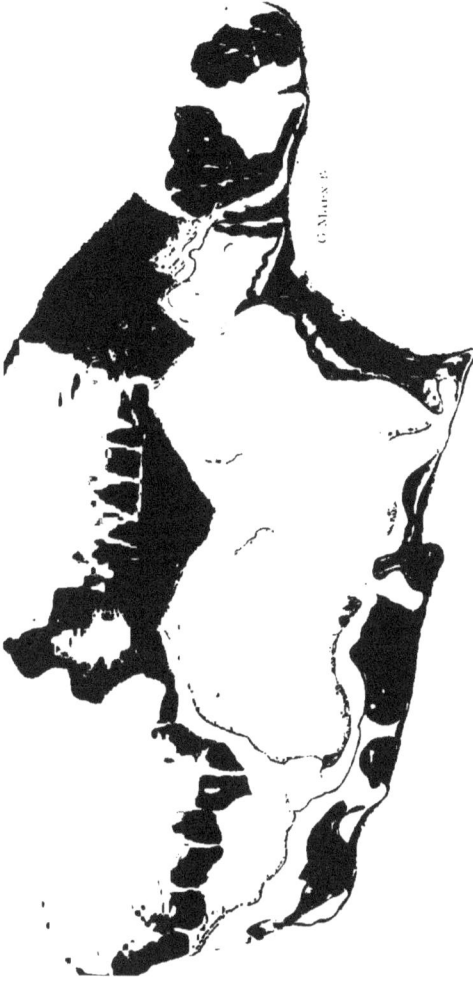

SWINE PLAGUE (Right Lung)

SWINE PLAGUE. (Right Lung.)

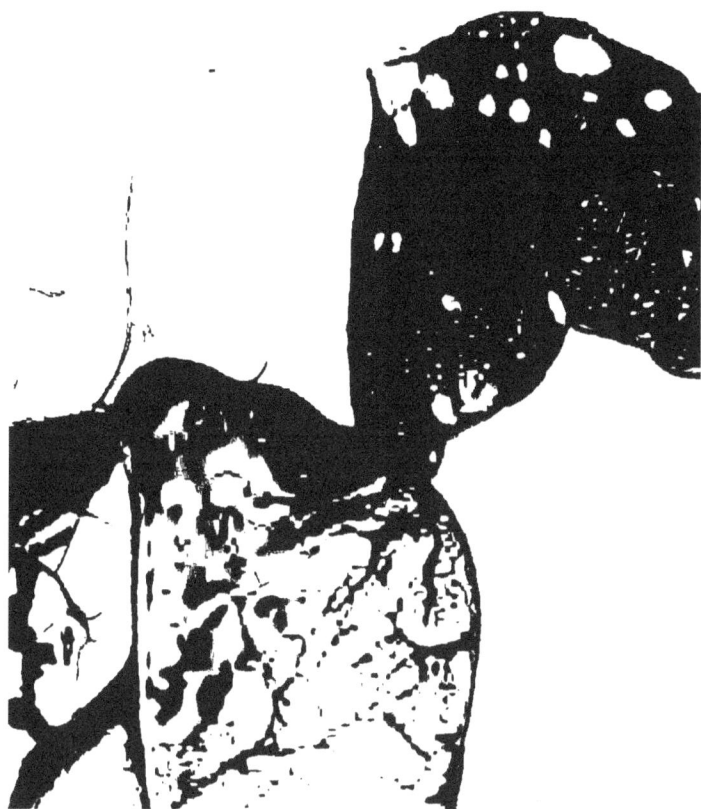

Fig. 1

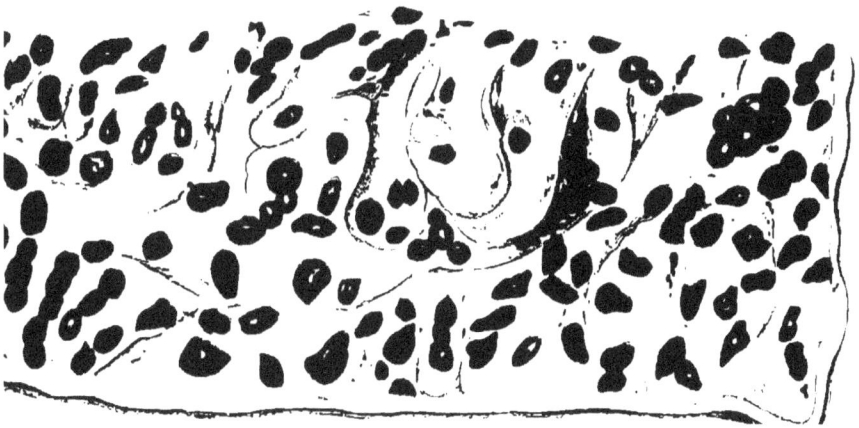

Fig. 2

Fig 1

Fig 2

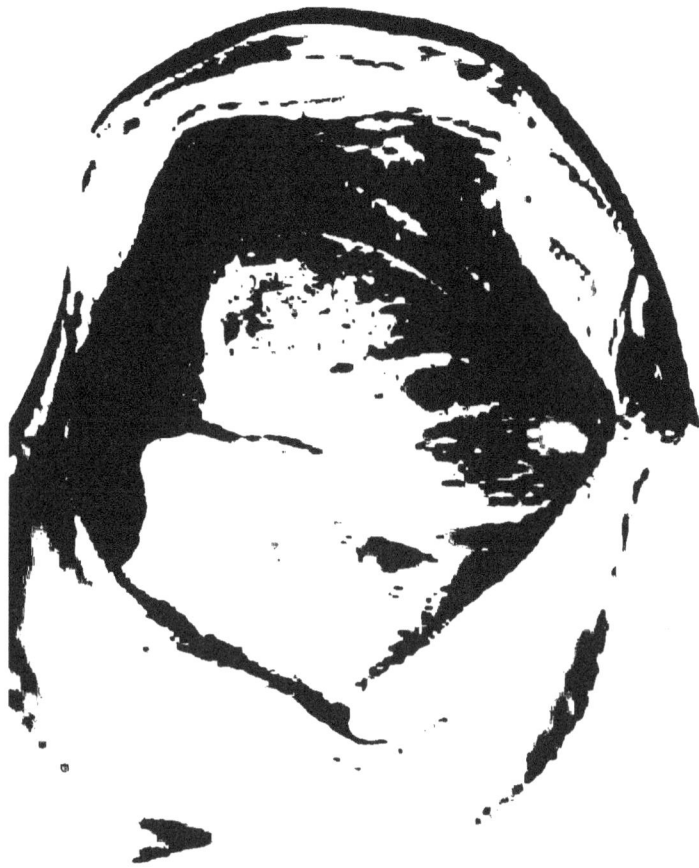

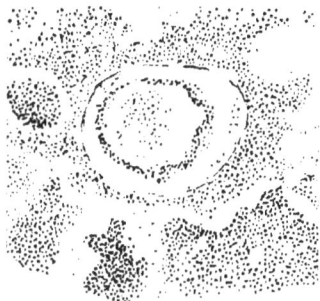

INDEX.